Introduction to Corrosion - Basics and Advances

Edited by Ambrish Singh

Published in London, United Kingdom

Introduction to Corrosion - Basics and Advances
http://dx.doi.org/10.5772/intechopen.104001
Edited by Ambrish Singh

Contributors
Nadjib Chafai, Khalissa Benbouguerra, Malak Rehioui, Sajjad Akramian Zadeh, Lakha V. Chopda, Mohamed Refai, Gamal E.M. Nasr, Zeinab Abdel Hamid, Sameer Kumar Dohare, Bharat Chandra Sahu, Guowen Yao, Xuanbo He, Zengwei Guo, Jiawei Liu, Jiangshan Lu, Shahin Kharaji

First published in London, United Kingdom, 2023 by IntechOpen
IntechOpen is the global imprint of INTECHOPEN LIMITED, registered in England and Wales, registration number: 11086078, 5 Princes Gate Court, London, SW7 2QJ, United Kingdom

British Library Cataloguing-in-Publication Data
A catalogue record for this book is available from the British Library

Additional hard and PDF copies can be obtained from orders@intechopen.com

Introduction to Corrosion - Basics and Advances
Edited by Ambrish Singh
p. cm.
Print ISBN 978-1-83768-667-4
Online ISBN 978-1-83768-668-1
eBook (PDF) ISBN 978-1-83768-669-8

Meet the editor

Dr. Ambrish Singh is a professor in the Department of Chemistry, Nagaland University, Lumami, Nagaland, India. Previously, he worked in the School of New Energy and Materials, Southwest Petroleum University, China for 9 years. Dr. Singh is a Fellow of the Royal Society of Chemistry (FRSC). His research interests include corrosion, electrochemistry, green chemistry, quantum chemistry, smart coatings, polymers, nanomaterials, composites, and petroleum engineering. He is the recipient of several awards, including the prestigious Sichuan 1000 Talent Award from the Sichuan government, China for his outstanding research contributions as a faculty, the President's Award for exceptional post-doctoral research work, and the Young Scientist Award from the Council of Science and Technology (UPCST), Lucknow, India. He has published more than 160 SCI peer-reviewed research papers in high-impact journals. He is a reviewer for more than 43 journals and an editor for several others. Dr. Singh has been invited to present his work at several national and international conferences, seminars, and workshops. He is the author/editor of several book chapters and books. He has drafted five patents based on his new and innovative findings and was awarded one patent in China. He has finished several state and provincial projects in China and India. He was a consultant to King Fahd University of Petroleum & Minerals (KFUPM), Saudi Arabia for an international project. He is a member of the Indian Science Congress Association (ISCA), The Association for Materials Protection and Performance (AMPP), Society for Petroleum Engineers (SPE), and American Chemical Society (ACS).

Contents

Preface

Although it is frequently silent and subtle, corrosion is perhaps the most important factor contributing to the decline of social systems. One of the most important technological challenges facing our country and the rest of the globe is extending the life of structures and engineered materials while maintaining public safety and the environment. A wide range of materials and systems are affected by corrosion-related issues, which have an impact on our daily life. Numerous economic sectors in the country are severely impacted by corrosion. Corrosion costs billions of dollars every year in lost revenue. By utilizing already accessible corrosion management technology, the annual cost of corrosion can be cut to 20 to 25 percent of that amount. Materials selection, correct design, electrochemical protective coating, and inhibitors are common corrosion control methods.

We are subjected to many types of corrosion or degradation daily. Corrosion is a risky and very expensive issue. It can cause bridges and buildings to fall, oil pipelines and chemical factories to leak, and restrooms to flood. Corroded medical implants may result in blood poisoning, corroded electrical contacts can result in fires and other issues, and air pollution has resulted in corrosion damage to works of art all over the world. Corrosion is an interdisciplinary subject and is a burning topic for researchers all over the world. It is equally related to chemistry, chemical engineering, mechanical engineering, aviation, metallurgy, materials science, petroleum, refineries, and many more.

Electrochemical processes are the main causes of the most prevalent types of corrosion. General corrosion happens when the majority of or all the atoms on a metal surface are oxidized, causing the surface to corrode completely. Most metals are easily oxidized, which means they frequently lose electrons to oxygen (among other elements) in the air or in water. Oxygen joins with the metal to produce an oxide as it is reduced (gains electrons). Corrosion is a sort of redox reaction where oxidation and reduction take place simultaneously.

The existing corrosion protection technologies are unable to stop corrosion. However, they can mitigate corrosion to some extent. Utilizing inhibitors, coatings, and other corrosion prevention techniques to solve this issue is a successful strategy. Furthermore, the creation of biodegradable and environmentally friendly inhibitors is becoming a crucial issue as regulations for the use of inhibitors get stricter and more demanding in terms of the ecological aspect. To prevent the corrosion of metals and alloys, a variety of techniques are used, including those that utilize polymers, nanocomposites, biopolymers, plant extracts, and many more.

I would like to thank all the authors and reviewers for their invaluable contributions.

Dr. Ambrish Singh (FRSC)
Professor,
Department of Chemistry,
Nagaland University,
Lumami, Zunheboto, Nagaland, India

Section 1

Corrosion Inhibitots

Chapter 1

Small Organic Molecule as Corrosion Inhibitors for Mitigating Metal Corrosion

Lakha V. Chopda

Abstract

Metal corrosion constitutes degradation of metals in the presence of favorable corrosive atmosphere. It worsens metal quality. The prevention of metal corrosion is so significant to save metals for their better utility. Corrosion inhibitors are widely used for the mitigation of metal corrosion. Small organic molecules as corrosion inhibitors are showed prominent corrosion inhibitive property because of their unique electron donating capacity to the metal orbitals. The bonding occurred between organic molecules and metals are main aspect to retard the corrosive environment toward metal.

Keywords: metal, corrosion, organic molecule, inhibitor and prevention

1. Introduction

Corrosion is the naturally happening process which transfers metals into their stable form. Metals easily facilitated to the corrosion in the presence of corrosive media (acidic, basic, and brine). In other words, it is the gradual deterioration of metals in the existence of favorable reactive environment. Nowdays, corrosion of metal became global problem as it worsened the quality of metals and directly or indirectly affect economy of any country. The control of corrosion metals using appropriate methodology is so important. Application of corrosion inhibitor for the prevention of corrosion found prominent attention. The suitable concentration of corrosion inhibitor reduced the corrosion rate without altering the concentration of corrosive media [1]. For the control of metal corrosion, various kinds of inhibitors including inorganic compounds, plant extract, and organic molecules are commenced for the mitigation of metal corrosion as illustrated in **Figure 1** [2]. Small organic inhibitors are widely applied for the control of metal corrosion as they are easy to prepare compared to complex organic molecules [3]. The best example of small organic molecule as corrosion inhibitor is a BTA [4]. BTA effectively mitigated metal corrosion. The presence of unique functionality in the small organic compounds is strongly react with metal by chemical or physical bonding and preserved metal against harsh corrosive media. Corrosion inhibitors interact with metal at metal/solution interface by forming film on the metal surface and hindered corrosion reaction. The formed protective

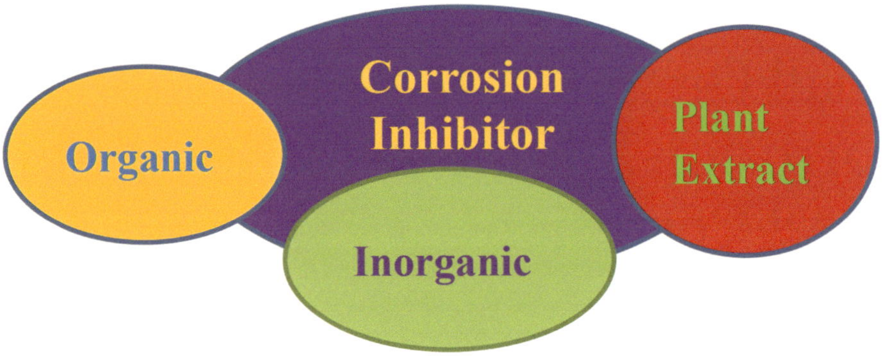

Figure 1.
Classification of corrosion inhibitor.

film limits corrosive media, oxygen, and water diffusion toward metal surface, so reducing corrosion rate. Any type of corrosion inhibitors can be categorized into cathodic, anodic, or mixed type which depending on their influence in the suppressing cathodic/anodic reaction or both [5]. This chapter includes the application of small organic molecule for the prevention of corrosion of various metals in different corrosive media. This chapter includes the application of small organic molecule for the prevention of corrosion of various metals in different corrosive media.

2. Application of small organic molecule as corrosion inhibitors

Corrosion inhibition activity of small organic molecules such as imidazole (IM), 2-ethylimidazole (EI), thiophene (TH), 2-ethylthiophene (ET), 3-methoxythiophene (MT), pyridine (PY), 4-ethylpyridine (EP), and 4-methoxypyridine (MP) have been assessed for the mild steel (MS) in 1 M HCl [6]. The corrosion inhibition property of inhibitors has been determined by the potentiodynamic polarization (PDP), linear polarization resistance (LPR), and electrochemical impedance spectroscopy (EIS) methods. The parameters such as a decrease in the corrosion current density (I_{corr}) values and an increase in the polarization resistance (Rp) revealed corrosion

Sr. No.	Corrosion inhibitor	Corrosion inhibition efficiency		
		PDP	LPR	EIS
1	Imidazole (IM)	70.4%	78%	80.1%
2	2-ethylimidazole (EI)	74.3%	81.1%	84.3%
3	Thiophene (TH)	59.9%	63.7%	64.2%
4	2-ethylthiophene (ET)	22.3%	52%	53.4%
5	3-methoxythiophene (MT)	58.1%	82.6%	82.8%
6	Pyridine (PY)	74.4%	82.9%	67.6%
7	4-ethylpyridine (EP)	76.9%	88.7%	85.9%
8	4-methoxypyridine (MP)	56%	67.5%	71.5%

Table 1.
Corrosion inhibition efficiency of eight heterocycles at 10 mM.

inhibition action of inhibitors. The corrosion inhibition efficiency increased with enhancing of the concentration of inhibitors. The inhibitor EP showed highest corrosion inhibition efficiency. The EP displayed 85.9%, 76.9%, and 88.7% corrosion inhibition efficiency as determined by PDP, LPR, and EIS. The PDP method indicated that corrosion inhibitors behaved as mixed type of inhibitors (control on the anodic and cathodic reaction). The corrosion inhibition efficiency of all eight compounds is displayed in **Table 1**.

The organic compounds (Z) -4 - ((2-bromobenzylidene) amino) -5-methyl-2-4-dihydro -3H-1,2,4-triazole-3-thione (**a**) and (Z) -4 - ((3-bromobenzylidene) amino) -5-methyl-2-4-dihydro-3H-1,2,4 -triazole-3-thione (**b**) have investigated as corrosion inhibitors for MS in 1 M HCl [7]. The PDP method showed that as concentration of both inhibitors increased led to enhance the corrosion inhibition efficiency. The compounds **a** and **b** showed 83.66% and 82.84% corrosion inhibition efficiencies at 10^{-3} M concentration. The same trend has been observed by EIS method. The increase in the value of charge transfer resistance (R_s) and increase in the value of double-layer capacitance (C_{dl}) with increased inhibitor concentration proved the corrosion inhibition action of inhibitor. Both inhibitors **a** and **b** showed 89.51% and 84.5% corrosion inhibition efficiencies at 10^{-3} M concentration. Both inhibitors followed the Langmuir adsorption isotherm and value of free energy (ΔG^0) is negative revealed that adsorption is spontaneous as well inhibitors adsorbed over metal surface by chemisorption. The effect of temperature showed that as temperature increased from 45 to 75°C, it causes a decrease in the corrosion inhibition efficiency. The compound 3-methyl-6-oxo-4-(thiophen-2-yl)-4,5,6,7-tetrahydro-2H-pyrazolo[3,4-b]pyridine-5-carbonitrile (TPP) displayed remarkable corrosion inhibition efficiency for MS in 1 M HCl [8]. The corrosion inhibition action of inhibitor increased as the concentration of inhibitor increased from 50 ppm to 200 ppm. The weight lost method displayed 94.28% corrosion inhibition efficiency at 200 ppm, while 95.75% and 96.83% corrosion inhibition efficacies by inhibitor at same concentration as determined by PDP and EIS methods, respectively. The PDP method showed that inhibitor followed the mixed type of behavior with more control on the cathodic reaction. The Langmuir adsorption isotherm was best fitted, and SEM images indicated that homogeneous adsorption film formed by the inhibitor TPP on the metal surface. The corrosion inhibition action of two Schiff bases such as 3-((phenylimino)methyl)quinoline-2-thiol (PMQ) and 3-((5-methylthiazol-2-ylimino)methyl) quinoline-2-thiol (MMQT) on mild steel surface has investigated by quantum chemical calculation and molecular dynamics (MD) simulation methods [9]. Quantum chemical parameters such as E_{HOMO}, E_{LUMO}, energy gap (ΔE), dipole moment (μ), electronegativity (χ), global hardness (η), and fraction of electron transfers have been determined, and parameter such as local reactive sites of molecule have been analyzed by Fukui indices. The adsorption behavior of the inhibitor molecules on Fe (1 1 0) surface has been analyzed using molecular dynamics simulation. The binding strength of the concerned inhibitor molecules on mild steel surface follows the order MMQT>PMQ that was in good agreement with the experimentally determined inhibition efficiencies. Three oxazole derivatives (Compound 1: methyl-4-naphthalen-2-ylmethylene-4H-oxazol-5-one, Compound 2: 4-(2-methyl-5-oxo-oxazol-4-ylidenemethyl)-benzaldehyde, and Compound 3: 4-(2-methyl-5-oxo-oxazol-4-ylidenemethyl)-benzene) were used as corrosion inhibitors for carbon steel (CS) (API5LX60) in 1 N H_2SO_4. The corrosion inhibition actions of three inhibitors were studied by gravimetric (weight lost) method in the concentration range of 50–200 ppm [10]. The corrosion inhibition efficiency increased with increased in concentration from 50 to 200 ppm. The highest corrosion inhibition efficiency attained at 200 ppm. At 200 ppm, inhibitors **1**, **2**, and **3** displayed 90.7%, 78.31%, and 75.66% corrosion inhibition efficiencies, respectively. The inhibitor **1** showed highest corrosion inhibition efficiency. The same trend has been found by the electrical resistance measurement. The inhibitor **1** exhibited highest resistance (2.9 Ω), while inhibitors **2** and **3** displayed 2.8 and 2.3 Ω value of resistance, respectively. The corrosion prevention of three choline amino acid ionic liquids ([Ch][AA]ILs) was evaluated for mild steel in 1 M HCl in the temperature range of 298–328 K [11]. The result gained from PDP method indicated that [Ch][AA]ILs influenced the cathode and anode reaction. It showed that [Ch][AA]ILs displayed mixed-type inhibitors. Among the evaluated corrosion inhibitors, choline phenylalanine ionic liquids ([Ch][Phe]) showed the highest corrosion efficiency. The values of standard free energy of adsorption (ΔG_{ads}) lied in the range between -20 kJ·mol^{-1} and $-$ 40 kJ·mol^{-1} revealing that inhibitor adsorbed on the metal surface by physical and chemical adsorption mechanism. The inhibitors formed a thick monolayer adsorption layer on the metal surface which effectively preventing metal-corrosive media interaction. The molecular dynamics simulation findings showed that the corrosion inhibitor molecules replaced the solvent water or any other ions pre-adsorbed on the metal surface during the adsorption process and protecting the metal against corrosion. As a cheap natural biomolecule, glucose has proven significant corrosion inhibitor for various metals in the acidic and basic media [12]. The two thiocarbohydrazide-modified glucose derivatives (TBTD-1 and TBTD-2) were synthesized using N-glycosylic linkage. The both inhibitors TBTD-1 and TBTD-2 found to be potential green corrosion inhibitors for the prevention of corrosion of carbon steel pipelines in oil and gas industry. TBTD-1 and TBTD-2 displayed excellent corrosion inhibitive action with an inhibition efficiency of 99.1% for TBTD-1 and 99.4% for TBTD-2. Oleic imidazole (OIM) and mercaptoethanol (ME) are used as corrosion inhibitors for carbon steel in 3.5 wt% NaCl aqueous solution saturated by CO_2 at 60°C [13]. Both the WL and the electrochemistry method showed that the synergistic effect between OIM and ME significantly improved the corrosion inhibition performance of the mixture corrosion inhibitors. The inhibition efficiency of the mixture corrosion inhibitors is high up to 96.56% when the ratio of OIM and ME is 3 to 1. The quantum chemical parameters calculated at the B3LYP/6–311++G(d,p) level displayed that OIM can be preferentially adsorbed for its low energy gap, while molecular dynamic simulation results showed that OIM is preferentially adsorbed onto the Fe(110) surface with a adsorption energy of -1583.7 kcal·mol^{-1}. The adsorbed OIM molecules formed a film with voids and ME molecules fill in the voids forming a bilayer film. The entrapping of inhibitor directly inside the intrinsically conducting polymer (ICP) matrix of polypyrrole (PPy) [14]. The material was coated by using electrodeposited in the presence of β-cyclodextrine, benzotriazole, and 8-hydroxyquinoline in the deposition of electrolyte-bearing pyrrole and 3-nitrosalicylate as counter-anion. The self-healing performance was investigated by monitoring both the corrosion potential in an electrolyte-filled defect and the delamination behavior by scanning kelvin probe (SKP). An excellent performance in terms of an extraordinarily significant passivation effect is found. The hollow mesoporous organosilica nanoparticles (HMONs) encapsulated by 2-mercaptobenzothiazole (MBT) corrosion inhibitor are used to prevent the corrosion of Cu in 3.5 wt% NaCl [15]. MBT@HMON was well characterized, and the release kinetics of MBT from HMON under different dithiothreitol (DTT) concentration and pH conditions was studied. The WL and EIS were used to study the controlled release of corrosion inhibitor MBT from HMON by tuning the redox and pH conditions of corrosion medium. The 1.6 $g \cdot L^{-1}$ concentration of MBT@HMON was added in the corrosion media which effectively prevented the corrosion of copper alloy. Two modified polyaspartic acid (PASP) derivatives poly(cysteaminoaspartamide) and poly(methionenoaspartamide) were successfully demonstrated corrosion inhibitors for C1018 steel in 3.5% NaCl-saturated CO_2 brine solution by PDP, LPR, and EIS methods [16]. SEM images revealed that uniformed layer of inhibitors formed on the metal surface. Anticorrosion effect of tetracycline drug on carbon steel alloy (ck45) in 1 M HCl using PDP and EIS method was demonstrated [17]. Using 300 ppm concentration of the drug showed 82% and 78% corrosion inhibition efficiencies as determined by PDP and EIS methods, respectively. The PDP measurement showed that tetracycline drug behaved as mixed-type inhibitor. The adsorption of drug molecules on the alloy surface followed the Langmuir adsorption isotherm. $\Delta G°_{ads}$ valued suggested that the drug molecules adsorbed on the alloy surface through physisorption mechanism. Additionally optical microscopy and SEM images also confirmed corrosion inhibition action of drug by the formation of a protective film over the metal surface. Corrosion inhibition performance of imidazole drugs omeprazole (OMP) and its byproducts omeprazole sulfide (OMP-1) and omeprazole sulfonate (OMP-2) on Q235 steel in 1 M HCl was evaluated using WL, PDP, EIS, SEM, and XPS methods as well quantum chemical calculation and molecular dynamics simulation [18]. The obtained results showed that OMP, OMP-1, and OMP-2 are proven to be effective corrosion inhibitors for Q235 steel in 1 M HCl. The inhibition efficiency of above 95% was found at 298 K with 0.2 mM inhibitor concentration. The PDP method was revealed that inhibitors are mixed-type corrosion inhibitors by obeyed Langmuir adsorption isotherm. SEM and XPS studies further confirmed that three inhibitor molecules showed significant corrosion inhibition action by adsorbing on the surface of Q235 steel which formed barrier. Two heteroatom-containing phosphor amides, N-(5-methylisoxazol-3-yl)-P,P-diphenylphosphinic amide (MAPP) and diphenyl (5-methylisoxazol-3-yl)phosphoramidate (PMAP) were found to potent corrosion inhibitors for carbon steel in HCl [19]. The polyaspartic acid (PASP) derivatives modified with benzothiazole compounds (R-ABT: R = H, Me, and OMe) were demonstrated as fluorescent green corrosion inhibitors [20]. The corrosion performance of three inhibitors on carbon steel in cooling water was investigated using the electrochemical test, WL measurement, SEM, AFS, XPS, DFT, and MD simulation. The PDP method indicated that three derivatives are mixed-type inhibitors, and their corrosion inhibition efficiencies trend followed the order as: PASP-OMe-ABT (96.45%) > PASP-Me-ABT (91.38%) > PASP-H-ABT (89.38%). The three inhibitor molecules improved the corrosion of carbon steel surface by physisorption and chemisorption. Pectin obtained from the waste peel and residue of various plants material is used to synthesize environmentally friendly and biodegradable green corrosion inhibitor. Two corrosion inhibitors of pectin modified by glycine(Gly) and lysine (Lys) were prepared and applied for the control of corrosion of N80 carbon steel in CO_2-saturated groundwater [21]. The 6.26% and 7.20% of gly and lys were amidated to pectin as confirmed by elemental analysis, FTIR spectroscopy, ^1H NMR, and XRD. Both inhibitors P-Gly (99.01%) and P-Lys (99.36%) showed excellent corrosion inhibition efficiency at 800 mg/L for 120 h. Theoretical calculations done by QC based on DFT and GFN-xTB methods and MD-based RDFs revealed the chemisorption of the inhibitors. N-substituted methyl ethylenediamine derivatives, tetramethylethylenediamine (TDA), pentamethyldiethylenetriamine (PTA), and hexamethyltriethylenetetramine (HTA) are used as corrosion inhibitors for 20# steel in 1 M HCl [22]. The corrosion property of three inhibitors is conducted by WL, PDP,

EIS, and surface analysis techniques. Three methyl ethylenediamine derivatives displayed excellent inhibition properties. The order of inhibition efficiency of these inhibitors was TDA < PTA < HTA. PDP techniques revealed that three inhibitors act as mixed-type inhibitors. The adsorption mechanism of corrosion inhibitors followed the Langmuir isotherm and adsorption followed physisorption and chemisorption. Quantum chemical calculations and MD simulations well supported the experimental results. The excellent corrosion inhibition property of inhibitors was linked to the presence of tertiary amine groups which effectively bound with Fe. The corrosion inhibitor N1-(2-aminoethyl)-N2-(2-(2-(furan-2-yl)-4,5-dihydro-1H-imidazol-1-yl) ethyl) ethane-1,2-diamine (NNED) showed excellent corrosion inhibitive action for carbon steel in 1 M HCl [23]. It exhibited more than 90% corrosion inhibition efficiency at 5 ppm. The adsorption of NNED on the steel surface was spontaneous and followed the Langmuir adsorption isotherm. Xanthium cocklebur corrosion inhibitor effectively suppressed corrosion of carbon steel [24]. It effectively decreased the corrosion current and increased the corrosion potential. The adsorption of inhibitor confirmed the Frumkin adsorption isotherm model. The corrosion inhibition mechanism was attributed to the phenolic hydroxyl oxygen of 1,5-dicaffeoylquinic acid hybridizes that formed strong covalent bond with empty 3d orbital of Fe and hydrogen bond formed between hydrogen of hydroxyl with oxygen of hydroxyl presence on the surface of γ-FeOOH (0 1 0). Diaminodecane-functionalized graphene oxide (DAD-GO) and diaminododecane-functionalized graphene oxide (DADD-GO) were demonstrated as corrosion inhibitors for carbon steel corrosion in 15.0% HCl [25]. The inhibition activity enhanced the concentration at room temperature and obtained maximum to 84% for DAD-GO and 78% for DADD-GO at a concentration of 5 ppm for both. The increased temperature contributed to decrease in the corrosion inhibition efficiency. The Langmuir adsorption was followed by the both inhibitors, and corrosion properties of both inhibitors are well supported by AFM, SEM, EDX, and FTIR. All investigated methods revealed that functionalized both GO inhibitors formed protective layer after 24 h of immersion time. 3,5-bis(4-methylphenyl)-4-amino-1,2,4-triazole (MeAT) and 3,5-bis(4-methoxyphenyl)-4-amino-1,2,4-triazole (MxAT) were investigated as corrosion inhibitors for Cu in 1 M HCl [26]. 1,2,4-triazole derivatives are emerged as efficient corrosion inhibitors for Cu in 1 M HCl. The compound MxAT showed highest corrosion inhibition efficiency (87.9%) at 10^{-3} M concentration of MxAT at 303 K. The adsorption of MxAT molecule followed Langmuir adsorption isotherm and founded ΔG^{o}_{ads} indicated that adsorption of MxAT is chemisorption. 3--amino-1,2,4-triazole-5-thiol (ATAT) inhibited the corrosion of Al [27] in naturally aerated stagnant Arabian Gulf water. The corrosion inhibition action of ATAT was evaluated by EIS, cyclic potentiodynamic polarization (CPP), and potentiostatic current–time (CCT) measurements. EIS method showed that ATAT reduced corrosion of Al by enhancing the solution and polarization resistances, the impedance of the interface, and the maximum degree of phase angle. The other two techniques CPP and CCT revealed the corrosion inhibition property of ATAT by decreased of cathodic, anodic, and corrosion currents and corrosion rate and increased polarization resistance of Al. The pH-sensitive corrosion inhibitor benzotriazole (BTA) was capsulated on the surface of mesoporous silica with metal phenol network. A PH-triggered release of BTA from mesoporous silica as confirmed by UV-vis measurements proved that outstanding protection of Cu [28]. The anticorrosive activities of furfural derivatives prepared by ex situ and in situ are used to prevent the corrosion of mild steel immersed in CO_2-saturated water by WL, PDP, EIS, XPS, and quantum chemical calculation [29]. Furfural thiosemicarbazone (prepared by ex situ) showed inhibition efficiency 90.2% for 72 h in immersion. The same molecule, synthesized in situ showed 82.2% corrosion inhibition efficiency for 24 h of immersion, but corrosion inhibition efficiency decreased over the time due to incomplete conversion of the precursors in furfural thiosemicarbazone. The corrosion activity of mild steel in 1 M HCl was evaluated in the presence of (E)-3-(5-bromo-2-hydroxystyryl)quinoxalin-2(1H)-one (FR178) and (E)-3-(5-fluoro-2-hydroxystyryl) quinoxalin-2(1H)-one (FR179) [30]. The organic inhibitors emerged as efficient corrosion inhibitors for mild steel. Increasing concentrations of inhibitors contributed to high inhibition efficiencies that were due to blocking of both sides of metal (anodic and cathodic). The inhibition efficiency found to be 90.9% and 91.5% for both inhibitors FR178 and FR179, respectively, at 10^{-3} M. The anticorrosion activities of FR178 and FR179 decreased to 82% and 84%, respectively, at 328 K. PDP and EIS methods showed that both inhibitors are the mixed type. The X-ray, XRD, and SEM-EDS revealed about the formation of protective layer on the metal surface. DFT and molecular dynamics (MD) simulations supported the anticorrosion activity of both inhibitors. Two novel organic aromatic thiophene derivatives, 5-(thiophen-2-yl) – 1 H-tetrazole (TET) and thiophene-2-carbaldehyde oxime (OXM), were examined for the corrosion protection of mild steel in 1 M HCl [31]. The two organic inhibitors have significant resistance to the effect of time of immersion and corrosion inhibition efficiency of inhibitors increased with time. The maximum corrosion inhibition efficiency attained 94% and 95% for OXM and TET, respectively, at 10^{-3} M. The maximum shift in E_{corr} values found to be 10–17 mV signified thiophene compounds are mixed-type (anodic/cathodic) inhibitors. SEM micrographs confirmed the formation of a protective layer adsorbed on the mild steel surface. The electrophilic/nucleophilic attacks (Fukui indices) on the metal surface are theoretically determined and strong supporter of findings obtained by DFT and MC simulations and electrochemical methods. Ammonium salts DDPC and DDQC showed superior corrosion inhibition property for X65 carbon steel in the sour brine solution at different temperatures [32], and concentration were studied by various methods (WL, electrochemical measurements, and surface analysis (SKP, AFM, SEM, EDs, and ATR-FTIR). The obtained results indicated that corrosion rate decreased with increasing inhibitor concentration, and the inhibition efficiency reached to 98%. The adsorption of inhibitors on the metal surface followed the Langmuir adsorption isotherm. Corrosion inhibition property of inhibitors did not alter at elevated temperatures. Vitamin B_{12} studied as corrosion inhibitor for the mild steel [33]. The PDP and EIS methods were used to assess the corrosion activity of vitamin B_{12}. EIS method indicated that in the presence of vitamin B_{12}, there is an increase in the R_{ct} (charge transfer resistance) and a decrease in the value of capacitance layer (C_{dl}). It showed the formation of a protective layer over the mild steel surface. The surface morphology in the absence and presence of vitamin B_{12} on the mild steel was examined by FE-SEM. The smooth surface has been seen in the presence of inhibitor that revealed the protective behavior of vitamin B_{12}. Thiadiazole derivative, 1-phenyl-3-(5-thioxo-4,5-dihydro-1,3,4-thiadiazol-2-yl) thiourea (PTT), was synthesized with 5-amino-1,3,4-thiadiazole-2(3 H)-thione (ATT) [34]. The corrosion inhibitor PTT applied to control the corrosion of N80 carbon steel in supercritical CO_2. Corrosion inhibitor PTT showed the excellent corrosion inhibition efficiency (99.58%). DFT calculations confirmed the adsorption of PTT on steel surface by the formation of bonding with the three S atoms of (thione-thiadiazolely)thiourea fragment and with the benzene ring. Compared to ATT, PTT showed the significant improvement in the inhibition performance. It was due to the stronger adsorption of PTT by forming more S-Fe bonds and the extra C-Fe bonds. Corrosion protective property of coumarin-buta-1, 3-diene-conjugated dyes was investigated for steel in acidic environment [35]. The highest corrosion efficiency was estimated to be 97.62% as determined by PDP method. Amino acid-based surfactant sodium cocoyl glycinate (SCG) is studied as a corrosion inhibitor for mild steel 1 M HCl [36]. At 30°C, SCG showed 95.21% inhibition efficiency at 0.2864 mM. The corrosion inhibition efficiency was decreased to 73.21% at 60°C. Thermodynamic and kinetic parameters of MS corrosion and inhibitor adsorption were determined. The Gibbs free energy of adsorption (ΔG_{ads}) was calculated and found to be in the range of −30 and − 40 KJ mol^{-1}. It showed the mixed adsorption phenomena (chemisorption and physisorption). Sodium metamizole exhibited corrosion inhibitive action for carbon steel (CS) in 1 M HCl [37]. Corrosion inhibitor displayed 82.87% corrosion inhibition efficiency at 300 ppm and 25°C. The effect of temperature on the CS indicated that corrosion inhibition efficiency decreased at elevated temperature. It showed physisorption of inhibitor on the metal surface as supported by activation energy (80 kJ mol^{-1}) and free energy (20 kJ mol^{-1}). The adsorption of inhibitor followed Langmuir adsorption model. The PDP method indicated that inhibitor behaved as mixed type. EIS showed a decrease in the values of double-layer capacitance and an increase in the value of the charge transfer resistance with increased concentration of sodium metamizole. Schiff (1Z)-N-[2-(methylthio) phenyl]-2-oxopropane hydrazonoyl chloride (S1) showed anticorrosive property for mild steel in 1 M HCl [38]. Corrosion inhibition efficiency of S1 increased with increasing concentration of inhibitor and decreased with increase in temperature from 25° to 55°C. The maximum inhibition efficiency achieved is 87% at 2.5×10^{-3} M concentration at 25°C. Contact angle in the presence and absence of inhibitor was calculated. In the presence of inhibitor S1 showed 75°, while in the absence of inhibitor value of contact angle increased. Corrosion inhibitor followed the Langmuir isotherm. Cefotaxime sodium drug (Cefo) displayed corrosion inhibition property for Al in 0.1 M NaOH [39]. The corrosion inhibition efficacy increases with increasing concentrations of up to 300 ppm while decreasing with increase in temperature. The maximum inhibition efficiency of Cefo was 71.8% at 300 mg/L. Cefo adsorption on the Al surface is a mixed and spontaneous process that obeyed Freundlichs adsorption isotherm. XPS and FESEM/EDAX proved the presence of protective layers on the Al surface. Two bi-pyrazole-carbohydrazides, 1,1′-(propane-1,3-diyl)bis (5-methyl-1H-pyrazole-3-carbohydrazide) (P2PZ) and 1,1′-(oxy bis (ethane −2,1-diyl)) bis (5-methyl-1H-pyrazole-3- Carbohydrazide) (O2PZ) were evaluated as corrosion inhibitors for mild steel (MS) in 1.0 M HCl [40]. The corrosion inhibition efficiency achieved to 95% and 84% at 308 K and 10–3 mol/L of P2PZ and O2PZ, respectively. Corrosion inhibition efficiency increased with increase in concentration and decreased with increase in temperature. PDP method showed that P2PZ and O2PZ acted as mixed-type inhibitors and followed the Langmuir adsorption isotherm. The free energy of adsorption (ΔG_{ads}) values are found to be in the range of −39.94 to −37.36 KJ mol^{-1}. It confirmed the physicochemical adsorption process. SEM images indicated the formation of a protective layer on the surface of the metal in the presence of both inhibitors. Dopamine-modified polyaspartic acid (PASP-Dop) showed corrosion protective action toward mild steel [41]. Corrosion inhibition efficiency of PASP-Dop was estimated to be 90.9% at 298 K in 0.5 M H_2SO_4 using 100 ppm concentration of inhibitor. The corrosion inhibition efficiency of inhibitor did not alter at elevated temperature. 2-mercaptobenzothiazole (MBT) showed corrosion inhibition action for Zn in 1 M HCl [42]. PDP and EiS method showed an increase in charge transfer resistance (R_{ct}) and a decrease in corrosion current density (Icorr). The values of R_{ct} found to be 156.8, 225.8, 420.7, and 753.5 Ω cm^2 at 0, 100, 300 and 500 ppm MBT, and value of I_{corr} decreased as increase in the concentration of inhibitor. The order of I_{corr} is as follows: 500 ppm (89 μA cm^{-2}) < 300 ppm (230 μA cm^{-2}) < 100 ppm (340 μA cm^{-2}) < 0 ppm (955 μA cm^{-2}). Two xanthone derivatives, namely 12-(4-chlorophenyl)-9,9-dimethyl-8,9,10,12-tetrahydro-11H-benzo[a]xanthen-11-one (BX-Cl) and 9,9-dimethyl-12-(4-nitrophenyl)-8,9,10,12-tetrahydro-11H-benzo[a]xanthen-11-one (BX-NO2), were studied as corrosion inhibitor for P110 steel in 15% HCl [43]. The corrosion inhibition of BX-Cl and BX-NO_2 was improved in the presence of KI. WL method showed that corrosion inhibition efficiency increased from 62.10% to 92.21% for BX-Cl and 47.36 to 84.21% for BX-NO_2 as concentration increased from 50 to 200 mg/L. The corrosion inhibition efficiency of BX-Cl and BX-NO2 decreased with increase in temperature. EIS method indicated that R_{ct} values increased upon increasing the amount of BX-Cl and BX-NO_2 and reached to maximum value to 644.8 Ω cm^2. PDP confirmed the reduction in the I_{corr} value from 630 to 54 μA/cm^2.

3. Conclusion and future prospective

Corrosion of metal is the ambitious problem. The corrosion of metal can be controlled by using appropriate method. Corrosion inhibitors somewhat mitigated the metal corrosion. Small organic molecules played a significant role as corrosion inhibitors. The benefit to utilize small organic molecules is that they are easy to prepare (no multiple steps are involved) and exhibited good corrosion activity similar to complex molecules. There are numerous small molecules available The future direction in this filed is to apply small organic molecules which are most economical, less hazardous, and readily available from the natural sources and can show good anticorrosive property as well.

Author details

Lakha V. Chopda
Government Engineering College, Bhuj, Gujarat, India

*Address all correspondence to: lakhan.chopda@gmail.com

References

[1] Kadhum A, Al-Amiery AAH, Al-Azawi RJ. Corrosion inhibitors. A review. International Journal of Corrosion and Scale Inhibition. 2021;**10**:54-67. DOI: 10.17675/2305-6894-2021-10-1-3

[2] Wonnie Ma IA, Ammar S, Kumar SSA, Ramesh K, Ramesh S. A concise review on corrosion inhibitors: Types, mechanisms and electrochemical evaluation studies. Journal of Coating Technology Research. 2022;**19**:241-268

[3] Goyal M, Kumar S, Bahadur I, Verma C, Ebenso EE. Organic corrosion inhibitors for industrial cleaning of ferrous and non-ferrous metals in acidic solutions: A review. Journal of Molecualr Liquids. 2018;**265**:565-573. DOI: 10.1016/j.molliq.2018.02.045

[4] Desai PD, Pawar CB, Avhad MS, More AP. Corrosion inhibitors for carbon steel: A review. Vitenam Journal of Chemistry. 2022;**61**. DOI: 10.1002/vjch.202200111

[5] Aslam R, Mobin M, Zehra S, Aslam J. A comprehensive review of corrosion inhibitors employed to mitigate stainless steel corrosion in different environments. Journal of Molecular Liquids. 2022;**364**:119992. DOI: 10.1016/j.molliq.2022.119992

[6] Carranza MSS, Reyes YIA, Gonzales EC, Arcon DP, Franco FC. Electrochemical and quantum mechanical investigation of various small molecule organic compounds as corrosion inhibitors in mild steel. Heliyon. 2021;**7**:e07952. DOI: 10.1016/j.heliyon.2021.e07952

[7] Merimi I, EL Ouadi Y, Benkaddour R, Lgaz H, Messali M, Jeffali F, et al. Improving corrosion inhibition potentials using two triazole derivatives for mild steel in acidic medium: Experimental and theoretical studies. Materials Today: Proceedings. 2019;**13**:920-930. DOI: 10.1016/j.matpr.2019.04.056

[8] Singh P, Quraishi MA, Gupta SL, Dandia A. Investigation of the corrosion inhibition effect of 3-methyl-6-oxo-4-(thiophen-2-yl)-4,5,6,7-tetrahydro-2H-pyrazolo[3,4-b]pyridine-5-carbonitrile (TPP) on mild steel in hydrochloric acid. Journal of Taibah University for Science. 2016;**10**:139-147. DOI: 10.1016/j.jtusci.2015.07.005

[9] Kr S, Saha P, Ghosh A, Hens NC, Murmu PB. Density functional theory and molecular dynamics simulation study on corrosion inhibition performance of mild steel by mercapto-quinoline Schiff base corrosion inhibitor. Physica E: Low-Dimensional Systems and Nanostructures. 2015;**66**:332-341. DOI: 10.1016/j.physe.2014.10.035

[10] Burhagohain P, Sharma G, Bujarbaruah PM. Investigation of a few oxazolone molecules as corrosion inhibitor for API5LX60 steel in 1N H_2SO_4 solution. Egyptian Journal of Petroleum. 2022;**31**:37-45. DOI: 10.1016/j.ejpe.2022.06.006

[11] Li E, Li Y, Liu S, Yao P. Choline amino acid ionic liquids as green corrosion inhibitors of mild steel in acidic medium. Colloids and Surfaces A: Physiochemical and Engineering Aspects. 2023;**657**:130541. DOI: 10.1016/j.colsurfa.2022.130541

[12] Wang X, Lei Y, Jiang ZN, Zhang QH, Li YY, Liu HF, et al. Developing two thiocarbohydrazide modified glucose derivatives as high-efficiency green

corrosion inhibitors for carbon steel. Industrial Crops and Products. 2022;**188**:115680. DOI: 10.1016/j.indcrop.2022.115680

[13] Zhu Y, Qu S, Shen Y, Liu X, Lai N, Dai Z, et al. Investigation on the synergistic effects and mechanism of oleic imidazoline and mercaptoethanol corrosion inhibitors by experiment and molecular dynamic simulation. Journal of Molecular Structure. 2023;**1274**:134512. DOI: 10.1016/j.molstruc.2022.134512

[14] Yin Y, Prabhakar M, Ebbinghaus P, da Silva CC, Rohwerder M. Neutral inhibitor molecules entrapped into polypyrrole network for corrosion protection. Chemical Engineering Journal. 2022;**440**:135739. DOI: 10.1016/j.cej.2022.135739

[15] Li C, Zhao X, Meng C, Zhang T, Sun S, Hu S. Application of hollow mesoporous organosilica nanoparticles as pH and redox double stimuli-responsive nanocontainer in the controlled release of corrosion inhibitor molecules. Progress in Organic Coatings. 2021;**159**:106437. DOI: 10.1016/j.porgcoat.2021.106437

[16] Obot IB, Ul-Haq MI, Sorour AA, Alanazi NM, Al-Abeedi TM, Ali SA, et al. Modified-polyaspartic acid derivatives as effective corrosion inhibitor for C1018 steel in 3.5% NaCl saturated CO_2 brine solution. Journal of Taiwan Institute Chemistry Engineering. 2022;**135**:104393. DOI: 10.1016/j.jtice.2022.104393

[17] Shojaee S, Zandi MS, Rastakhiz N. The effect of tetracycline drug as a green corrosion inhibitor for carbon steel in HCl media. Journal of the Indian Chemical Society. 2022;**99**:100700. DOI: 10.1016/j.jics.2022.100700

[18] Liu Y, Wang Z, Chen X, Zhang Z, Wang B, Li H-J, et al. Synthesis and evaluation of omeprazole-based derivatives as eco-friendly corrosion inhibitors for Q235 steel in hydrochloric acid. Journal of Environment and Chemistry Engineering. 2022;**10**:108674. DOI: 10.1016/j.jece.2022.108674

[19] Gholivand K, Sarmadi-Babaee L, Faraghi M, Badalkhani-Khamseh F, Fallah N. Heteroatom-containing phosphoramides as carbon steel corrosion inhibitors: Density functional theory and molecular dynamics simulations. Chemical Physics Impact. 2022;**5**:100099. DOI: 10.1016/j.chphi.2022.100099

[20] Dou F, Han J, Li J, Zhang H, Qiao K, Kan J, et al. Exploration of novel polyaspartic acid derivatives as fluorescent eco-friendly corrosion inhibitors for the carbon steel: Electrochemical, surface analysis (SEM/XPS) and theoretical calculation. Colloids and Surfaces A: Physiochemical and Engineering Aspects. 2023;**658**:130606. DOI: 10.1016/j.colsurfa.2022.130606

[21] Wang D-Y, Wang J-H, Li H-J, Wu Y-C. Pectin-amino acid derivatives as highly efficient green inhibitors for the corrosion of N80 steel in CO2-saturated formation water. Industrial Crops and Products. 2022;**189**:115866. DOI: 10.1016/j.indcrop.2022.115866

[22] Yang L, Fan H, Yan R, Zhang J, Liu S, Huang X, et al. N-substituted methyl ethylenediamine derivatives as corrosion inhibitors for carbon steel in 1 M hydrochloride acid. Journal of Molecular Structure. 2022;**1270**:133975. DOI: 10.1016/j.molstruc.2022.133975

[23] Chen Z, Fadhil AA, Chen T, Khadom AA, Fu C, Fadhil NA. Green synthesis of corrosion inhibitor with biomass platform molecule: Gravimetrical, electrochemical,

morphological, and theoretical investigations. Journal of Molecular Liquids. 2021;**332**:115852. DOI: 10.1016/j.molliq.2021.115852

[24] Li Y, Xu W, Li H, Lai J. Corrosion inhibition mechanism of Xanthium sibiricum inhibitor and its comprehensive effect on concrete performance: Experimental analysis and theoretical calculation. Construction Building Materials. 2022;**348**:128672. DOI: 10.1016/j.conbuildmat.2022.128672

[25] Saleh TA, Haruna K, Alharbi B. Diaminoalkanes functionalized graphene oxide as corrosion inhibitors against carbon steel corrosion in simulated oil/gas well acidizing environment. Journal of Colloid and Interface Science. 2023;**630**:691-610. DOI: 10.1016/j.jcis.2022.10.054

[26] Belghiti ME, El Ouadi Y, Echihi S, Elmelouky A, Outada H, Karzazi Y, et al. Anticorrosive properties of two 3,5-disubstituted-4-amino-1,2,4-triazole derivatives on copper in hydrochloric acid environment: Ac impedance, thermodynamic and computational investigations. Surfaces and Interfaces. 2020;**21**:100692. DOI: 10.1016/j.surfin.2020.100692

[27] Sherif E-SM. Electrochemical investigations on the corrosion inhibition of aluminum by 3-amino-1,2,4-triazole-5-thiol in naturally aerated stagnant seawater. Journal of Industrial and Engineering Chemistry. 2013;**19**:1884-1889

[28] Guo C, Cao J, Chen Z. Core-shell mesoporous silica–metal–phenolic network microcapsule for the controlled release of corrosion inhibitor. Applied Surface Science. 2022;**605**:154747. DOI: 10.1016/j.apsusc.2022.154747

[29] da Cunha JN, Evangelista BVD, Xavier AV, da Silva TU, et al. Study of furfural derivatives as a possible green corrosion inhibitor for mild steel in CO_2-saturated formation water. Corrosion Science. 2023;**212**:110907. DOI: 10.1016/j.corsci.2022.110907

[30] Ech-Chebab A, Missioui M, Guo L, El Khouja O, et al. Evaluation of quinoxaline-2(1H)-one, derivatives as corrosion inhibitors for mild steel in 1.0 M acidic media: Electrochemistry, quantum calculations, dynamic simulations, and surface analysis. Chemical Physics Letters. 2022;**809**:140156. DOI: 10.1016/j.cplett.2022.140156

[31] Fernine Y, Arrousse N, Haldhar R, Raorane CJ, et al. Novel thiophene derivatives as eco-friendly corrosion inhibitors for mild steel in 1 M HCl solution: Characterization, electrochemical and computational (DFT and MC simulations) methods. Journal of Environment Chemistry and Engineering. 2022;**10**:108891. DOI: 10.1016/j.jece.2022.108891

[32] Iravani D, Esmaeili N, Berisha A, Akbarinezhad E, Alibadi MH. The quaternary ammonium salts as corrosion inhibitors for X65 carbon steel under sour environment in NACE 1D182 solution: Experimental and computational studies. Colloids and Surfaces A: Physiochemical and Engineering Aspects. 2023;**656**:130544. DOI: 10.1016/j.colsurfa.2022.130544

[33] Kesari P, Udayabhanu G. Investigation of vitamin B12 as a corrosion inhibitor for mild steel in HCl solution through gravimetric and electrochemical studies. Ain Shams Engineering Journal. 2023;**14**:101920. DOI: 10.1016/j.asej.2022.101920

[34] Li YY, Jiang ZN, Wang X, Zeng XO, Dong CF, Liu HF, et al. Developing a robust thiadiazole derivative corrosion inhibitor for dynamic supercritical CO_2

aqueous environment: Electrochemical tests and DFT calculations. Corrosion Science. 2022;**209**:110695. DOI: 10.1016/j.corsci.2022.110695

[35] Bedair MA, Elaryian HM, Bedair AH, Aboushahba RM, Fouda AE-AS. Novel coumarin-Buta-1,3-diene conjugated donor–acceptor systems as corrosion inhibitors for mild steel in 1.0 M HCl: Synthesis, electrochemical, computational and SRB biological resistivity. Inorganic Chemistry Communications. 2023;**148**:110304. DOI: 10.1016/j.inoche.2022.110304

[36] Ganjoo R, Sharma S, Thakur A, Assad H, et al. Experimental and theoretical study of sodium Cocoyl Glycinate as corrosion inhibitor for mild steel in hydrochloric acid medium. Journal of Molecular Liquids. 2022;**364**:119988. DOI: 10.1016/j.molliq.2022.119988

[37] Salem AM, Wahba AM, El-Hossiany A, Fouda AS. Experimental and computational chemical studies on the corrosion inhibitive properties of metamizole sodium pharmaceutical drug compound for CS in hydrochloric acid solutions. Journal of the Indian Chemical Society. 2022;**99**:100778. DOI: 10.1016/j.jics.2022.100778

[38] Ziouani A, Atia S, Hamani H, Douadi T, Al-Noaimi M, Gherraf N. Molecular dynamic simulation and experimental investigation on the synergistic mechanism and synergistic effect of (1Z) N [2 (methylthio) phenyl] 2oxopropanehydrazonoyl chloride (S 1) corrosion inhibitor on mild steel in acid medium1M HCl. Journal of the Indian Chemical Society. 2023;**100**:100832. DOI: 10.1016/j.jics.2022.100832

[39] Mohamedien HA, Kamal SM, Taha M, El-Deeb MM, El-Deen AG. Experimental and computational evaluations of cefotaxime sodium drug as an efficient and green corrosion inhibitor for aluminum in NaOH solution. Materials Chemistry and Physics. 2022;**290**:126546. DOI: 10.1016/j.matchemphys.2022.126546

[40] Cherrak K, Khamaysa OMA, Bidi H, El-Massaoudi M, et al. Performance evaluation of newly synthetized bi-pyrazole derivatives as corrosion inhibitors for mild steel in acid environment. Journal of Molecular Structure. 2022;**1261**:132925. DOI: 10.1016/j.molstruc.2022.132925

[41] Chai C, Xu Y, Xu Y, Liu S, Zhehang L. Dopamine-modified polyaspartic acid as a green corrosion inhibitor for mild steel in acid solution. European Polymer Journal. 2020;**137**:109946. DOI: 10.1016/j.eurpolymj.2020.109946

[42] Eduok U, Faye O, Szpunar J, Khaled M. CS_2 mediated synthesis of corrosion-inhibiting mercaptobenzothiazole molecule for industrial zinc: Experimental studies and molecular dynamic simulations. Journal of Molecualr Liquids. 2021;**324**:115129

[43] Singh A, Ansari KR, Bedi P, Pramanik T, Ali IH, Lin Y, et al. Understanding xanthone derivatives as novel and efficient corrosion inhibitors for P110 steel in acidizing fluid: Experimental and theoretical studies. Journal of Physics and Chemistry of Solids. 2023;**172**:111064. DOI: 10.1016/j.jpcs.2022.111064

Chapter 2

Phosphonates and Phosphonic Acids: New Promising Corrosion Inhibitors

Nadjib Chafai and Khalissa Benbouguerra

Abstract

In this chapter we present our published research results concerning the use of phosphonates and phosphonic acids synthesized in our laboratory as corrosion inhibitors. Firstly, the corresponding synthetic pathways used to prepare this type of inhibitors have been illustrated. Also, the different experimental methods used to evaluate the inhibition activities of these derivatives have been presented in this chapter such as weight loss measurements, polarization curves, electrochemical impedance spectroscopy (EIS), scanning electron microscopy (SEM), atomic force microscopy (AFM), etc. On the other hand, the theoretical approaches such as Density Functional Theory (DFT) and Molecular Dynamic Simulations (MDS) are also implanted in this chapter in order to determine correlations between the experimental efficiencies and some calculated structural and electronic properties.

Keywords: phosphonates, phosphonic acids, synthesis, corrosion inhibitors, mechanism, DFT, MDS, SEM, AFM, adsorption

1. Introduction

The corrosion of metallic materials is one of the most common phenomena that cause problems and losses in billions of dollars annually in the industrial fields. In this context, several methods have been developed and applied to combat this phenomenon such as the use of corrosion inhibitors, cathodic protection, protective coating, and galvanization.

In recent decades, the use of the organic and inorganic corrosion inhibitors takes a great importance in the protection of metals against corrosion in various media. Generally, the mechanism of the action of these inhibitors consists to form an adsorptive protective layer on the metallic surface. Also, the adsorbed inhibitors can be connected to the metallic surface by means of chemical bonds (chemical adsorption) or physical forces (physical adsorption).

Recently, phosphonates and phosphonic acids are largely employed as effective corrosion inhibitors to protect metals against corrosion in various media [1–4]. Based on this, several pathways and procedures have been developed to prepare theses derivatives in good yields such as Michaelis–Arbuzov reaction [5, 6], Kabachnik–Fields reaction [7, 8], Pudovik reaction [9], Abramov reaction [10] and

Moedritzer-Irani reaction [11]. In most cases these synthetic reactions require the use of a catalyst or, microwaves or ultrasounds in order to improve their yields and to minimize the reaction time [12–15].

Generally, the inhibition activity of organic and inorganic compounds can be studied experimentally using several chemical, electrochemical and microscopic methods such as weight loss measurements, polarization curves, electrochemical impedance spectroscopy (EIS), scanning electron microscopy (SEM), atomic force microscopy (AFM), infrared spectroscopy (IR), ... etc. Moreover, the theoretical methods such as Density Functional Theory (DFT) and Molecular Dynamic Simulations (MDS) are effectively used in the field of corrosion inhibition. Generally, the DFT method can be used to correlate the experimental inhibition efficiencies with some structural and electronic parameters of the investigated inhibitors. On the other hand, the MDS are used to determine the adsorption modes of the inhibitive molecules on the metallic surface and calculating their adsorption energies.

The main objective of this chapter is to present of our published research results concerning the synthesis and the use of phosphonates and phosphonic acids as a new generation of corrosion inhibitors. All inhibitors and their anticorrosion results presented in this chapter are exclusively studied and provided by our research team in the Laboratory of Electrochemistry of Molecular Materials and Complex (LEMMC) at Ferhat ABBAS Setif-1 University, Algeria.

2. Synthetic pathways of phosphonates and phosphonic acids

Many synthetic pathways of phosphonates and phosphonic acids have been developed and presented in the literature. In this context, the most common and important of these pathways are discussed below:

2.1 Michaelis: *Arbuzov* reaction

This reaction is considered as the most used way to synthesize phosphonates derivatives. It consists of adding of a trialkylephosphite to an alkyl halide [5, 6]. **Figure 1** shows the path followed by this reaction.

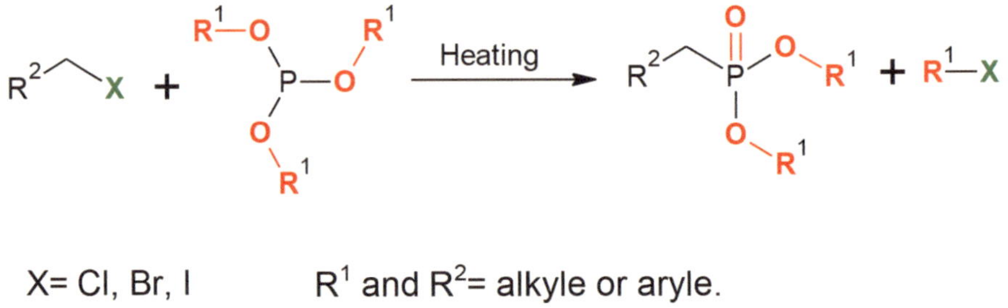

Figure 1.
Michaelis-Arbuzov reaction.

2.2 Kabachnik-Fields reaction (phospha-Mannich)

The Kabachnik-Fields reaction is one of the most important reactions for synthesizing α-aminophosphonates and α-aminophosphonic acids (**Figure 2**). In this

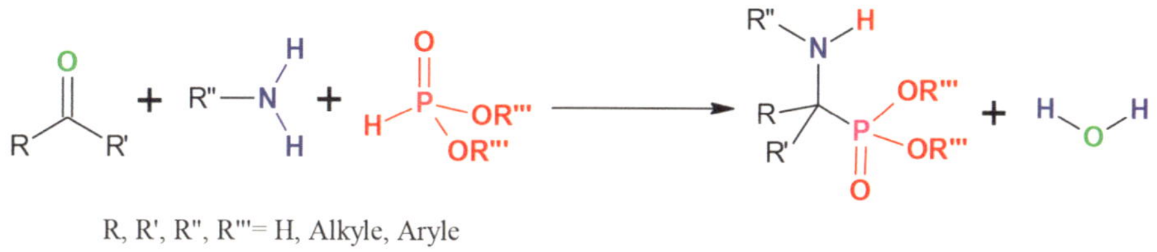

Figure 2.
Kabachnik-Fields reaction.

reaction, we react an amine, a carbonyl derivative and a dialkylephosphite together in one pot (one pot multi-component reaction) [7, 8].

2.3 *Pudovik* reaction

In this reaction, dialkylephosphites are converted to α-hydroxyphosphonates in the presence of carbonyl derivatives in basic medium [9]. The sequence of this reaction is given in **Figure 3**.

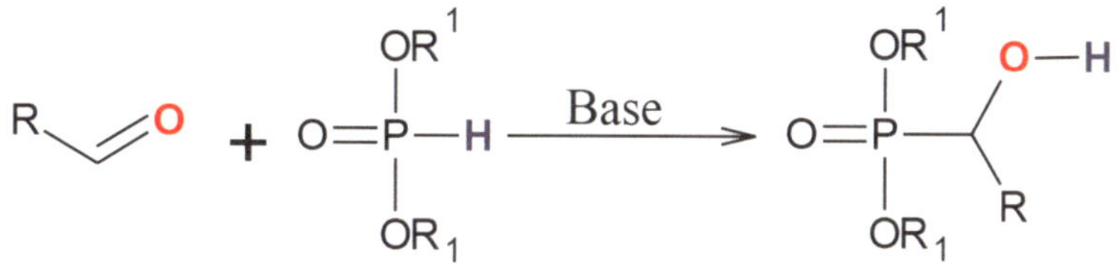

Figure 3.
Pudovik reaction.

2.4 *Abramov* reaction

Concerning the Abramov reaction, the trialkylephosphite rich in electron can undergo a nucleophilic addition to the carbon atom of the carbonyl compound [10]. The **Figure 4** represents the sequence of this reaction.

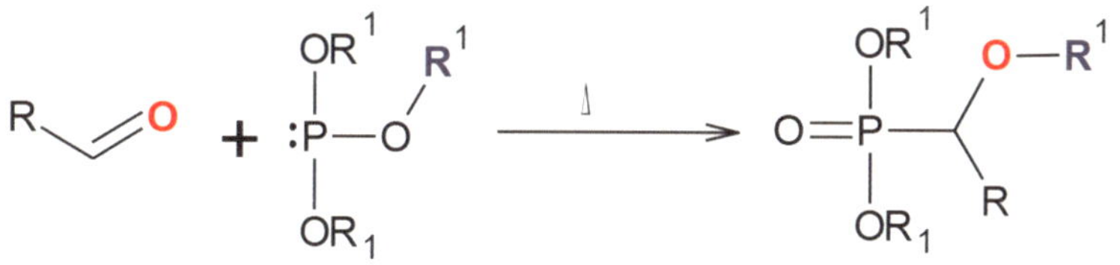

Figure 4.
Abramov reaction.

2.5 Moedritzer-*Irani reaction*

This reaction is developed by Kurt Moedritzer and Riyad R. Irani [11]. It is a new simple and direct procedure to prepare α-aminophosphonic acids basing on the Kabachnik-Fields and Mannich reactions. Generally, the principle of this reaction consists to react the phosphors acid and formaldehyde with primary or secondary amines (**Figure 5**).

$$R_{3-n}NH_n + nCH_2O + nHP(O)(OH)_2 \longrightarrow R_{3-n}N[CH_2P(O)(OH)_2]_n + nH_2O$$

n=1, 2 or 3

Figure 5.
Moedritzer-Irani reaction.

3. Experimental methods for the evaluation of the inhibition activity of phosphonates and phosphonic acids

Many experimental techniques and theoretical methods have been used to evaluate the corrosion inhibition activity of phosphonates and phosphonic acids. In this context, the most common and important of these techniques and methods are discussed below:

3.1 Weight loss measurements

The weight loss method is simple to implement and does not require significant equipment. Generally, the corrosion rate is determined after 24 h of immersion at a constant temperature equal to 25°C. In general, the operating protocol of this method consists in the first time the preparation of the metallic specimens used in weight loss tests. Then, we weigh the specimens before immersing them in the tested solutions. Also, each specimen was submerged in the tested solutions at constant temperature in absence and in presence of various concentrations of phosphonic inhibitors for a time of 24 h. After the expiration of the immersion time, the specimens have been recuperated from the tested solutions and rinsed with bi-distilled water. Finally, specimens were dried and weighed again.

The corrosion rate, surface coverage and inhibition efficiency are calculated by the following formulas:

$$A_{corr} = \frac{\Delta W}{S \times t} = \frac{W_1 - W_2}{S \times t} \tag{1}$$

$$\theta = \frac{A^0_{corr} - A_{corr}}{A^0_{corr}} \tag{2}$$

$$E_W(\%) = \frac{A^0_{\mathrm{corr}} - A_{\mathrm{corr}}}{A^0_{\mathrm{corr}}} \times 100 \tag{3}$$

where:

W_1: The mass of specimen before immersion in the tested solution.

W_2: The mass of specimen after immersion in the tested solution.

S: The surface area of the specimen.

t: The immersion time of each test.

$A_{corr}°$: The corrosion rate in the absence of the phosphonate inhibitor.

A_{corr}: The corrosion rate in the presence of the phosphonate inhibitor.

Accordingly, the weight loss method is largely applied to evaluate the corrosion inhibition activity of phosphonates and phosphonic acids. In this context, many our published works on the use of weight loss method indicates that the phosphonates and phosphonic acids are good inhibitors of the corrosion of steel in various aggressive

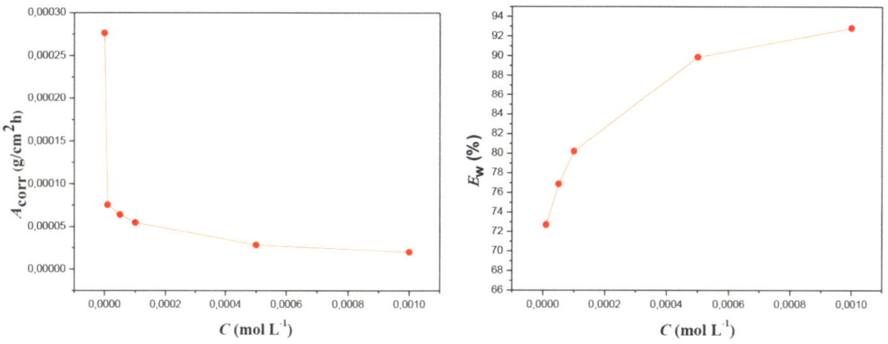

Figure 6.
Weight loss results for the inhibition activity of diethyl ((4-(dimethylamino)phenyl)(phenylamino)methyl) phosphonate obtained for the XC48 carbon steel in 0.5 Mol L^{-1} H_2SO_4 *at 25°C [16].*

media. **Figure 6** represents the weight loss results of the inhibition activity of diethyl ((4-(dimethylamino)phenyl)(phenylamino)methyl)phosphonate [16].

3.2 Polarization curves

The polarization curves technique is considered among the most widely used methods to determine the corrosion rate, the corrosion potential and the nature of the influence of the inhibitor on each of the elementary reactions, anodic and cathodic, at the electrode surface. Also, this method makes possible to determine the value of the corrosion current density by extrapolating the Tafel lines to the corrosion potential.

Generally, the following equation may be used to determine the inhibition efficiency obtained from the polarization curves ($E_p(\%)$):

$$E_p(\%) = \frac{\overset{\circ}{i}_{corr} - i_{corr(inh)}}{\overset{\circ}{i}_{corr}} \times 100 \tag{4}$$

where:

$\overset{\circ}{i}_{corr}$ and $i_{corr(inh)}$: are the values of the corrosion current density in the absence and in the presence of the inhibitor, respectively.

On the other hand, the values of surface coverage ratio (θ) can be calculated using the following equation:

$$\theta = \frac{i_{corr} - i_{corr(inh)}}{i_{corr}} \tag{5}$$

Concerning the application of the polarization curves technique in the evaluation of the corrosion inhibition activity of phosphonates and phosphonic acids, we observe from our previously published works that the majority of phosphonates and phosphonic acids derivatives inhibit corrosion by controlling the anodic and cathodic processes (mixed-type inhibitors) without affecting the dissolution of the metal in the anode or the evolution of hydrogen in the cathode [1]. Also, the adsorption of phosphonates and phosphonic acids on the metallic surfaces is responsible for the observed drop in $i_{corr(inh)}$ and the observed rise in $E_p(\%)$. As an example, **Figure 7**

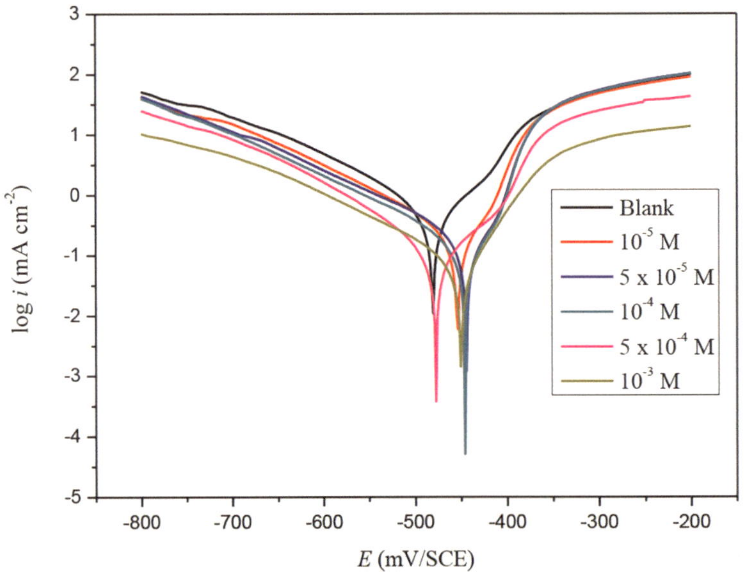

Figure 7.
Polarization curves for the inhibition activity of 4-(2-{[ethoxy(hydroxy)phosphonyl](3-nitrophenyl)methyl} hydrazinyl)benzoic acid obtained for the carbon steel in 0.5 Mol L^{-1} H_2SO_4 at 20°C [1].

represents the obtained polarization curves of the inhibition activity of 4-(2-{[ethoxy (hydroxy)phosphonyl](3-nitrophenyl)methyl}hydrazinyl)benzoic acid [1].

3.3 Electrochemical impedance spectroscopy (EIS)

In this technique, we measure the response the response of an electrode to a sinusoidal modulation of low amplitude of the potential as a function of the frequency. The strength of this technique is that it completely analyzes the mechanism of action of inhibitor on the metallic surface. So, the role of the inhibitor in the different processes occurring at the electrode such as charge transfer, diffusion, adsorption, etc., can be studied in detail, and values such as those of the transfer resistance and the polarization resistance can provide access to the measurement of the corrosion rate even in the case where the metal is covered with a protective layer [17]. Generally, to study the anticorrosion activity of phosphonates and phosphonic acids, the electrochemical impedance (EIS) measurements were performed around in the frequency range from 100 kHz to 10 mHz, with a signal of 5 mV sinusoidal amplitude. In this context, the inhibition efficiency (E_R(%)) can be calculated applying the electrochemical impedance spectroscopy results by using the following equation:

$$\eta_R(\%) = \frac{R_{t(inh)} - R_{t(0)}}{R_{t(inh)}} \times 100 \qquad (6)$$

Concerning the application of EIS study for the phosphonate derivatives inhibitors, we take as an example the Nyquist plot of [(2-hydroxy-5-methoxy-1,3-phenylene)bis (methylene)]bis(phosphonic acid) obtained for the carbon steel in 1 mol L^{-1} HCl at 25°C in the absence and in the presence of different concentrations of the inhibitor (**Figure 8**) [18]. Note that the diameter of the Nyquist diagram increases with the addition of the inhibitor, suggesting that the corrosion of carbon steel in acidic media is mainly controlled by a charge transfer process [19]. Also, it is clearly observed in **Figure 8** that the Nyquist diagrams of all tested concentrations present similar semi-circle shapes. This means that there is no significant change in the corrosion mechanism due to the addition of the inhibitor [20]. On the other hand, we observe that the obtained diagrams are not perfect semi-circles, because of the frequency dispersion which can be attributed to a surface heterogeneity which generates a frequency distribution. In general, this heterogeneity is due to the roughness of the surface and the chemical composition of carbon steel [21].

Also, the analysis of the results presented in **Figure 8** show that the charge transfer resistance (R_{ct}) values increase and the double layer capacitance (C_{dl}) values decrease with increasing the concentration of the phosphonate derivative. The increase in R_{ct} may be due to the formation of a protective film at the metal/solution interface. On the other hand, the decrease in C_{dl} values is due to the increase in the thickness of the electrical double layer, which indicates that the phosphonate derivative act by adsorption on the metal surface.

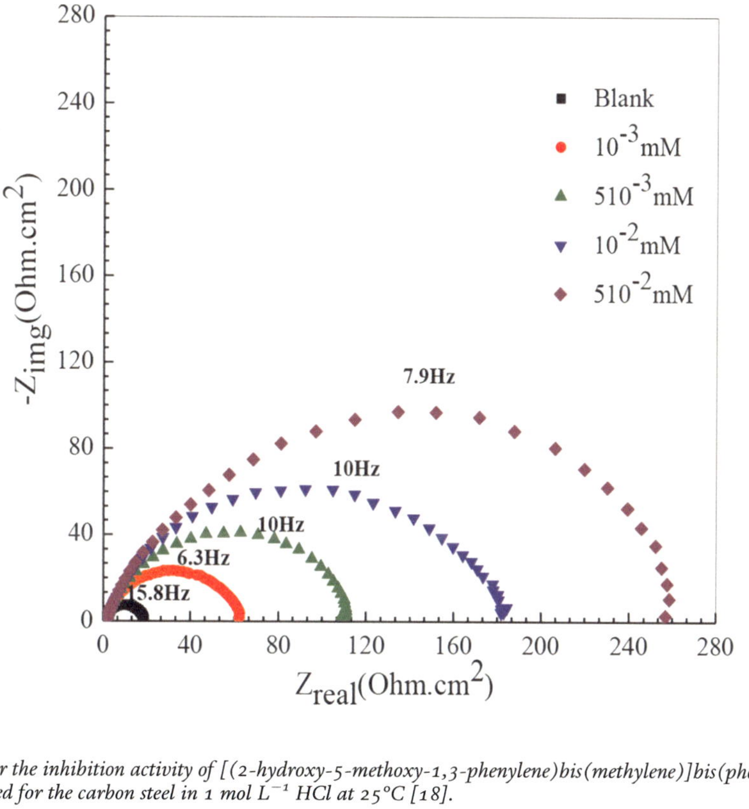

Figure 8.
EIS results for the inhibition activity of [(2-hydroxy-5-methoxy-1,3-phenylene)bis(methylene)]bis(phosphonic acid) obtained for the carbon steel in 1 mol L^{-1} HCl at 25°C [18].

3.4 Scanning electron microscopy (SEM)

The Scanning Electron Microscopy (SEM) technique is largely used in the corrosion inhibition field. The main objective of this technique is to visualize the surface morphology of metals in the absence and in the presence of inhibitory molecules. Briefly, a scanning electron microscope uses a very fine beam of electrons which scans, point by point, the surface of the sample to be observed. In this context, this technique allows researchers to visualize what is happening on the metal surface at the microscopic scale and to know what changes are made on the surface of the metal after the addition of the inhibitor (e.g. the addition of phosphonates and phosphonic acids inhibitors).

Figure 9 represents the SEM image of mild steel specimen in the presence of 10^{-3} mol/L of Ethyl hydrogen [(2-methoxyphenyl)(methylamino) methyl] phosphonate [22]. The examination of the obtained SEM image of the tested phosphonate compound shows an observed reduce in roughness of the surface of mild steel. So, we can say that the metallic surface has been protected against corrosion when the phosphonate derivatives have been added to the corrosive medium. This phenomenon is explained by the formation of an adsorbed layer (thin protective film) of the phosphonate derivative on the metal surface.

Figure 9.
SEM image of mild steel in the presence of 10^{-3} mol/L of ethyl hydrogen [(2-methoxyphenyl)(methylamino) methyl]phosphonate [22].

3.5 Atomic force microscopy (AFM)

Atomic force microscopy (AFM) is another useful microscopic technique which is extensively used in corrosion inhibition studies. Especially, this technique makes to determine the formation of a protective layer of the inhibitor on the metal surface by measuring the variation in the roughness values of the metal surface before and after the addition of the inhibitor molecule.

An example of the use of the AFM technique as a useful way to illustrate the formation of adsorbed layers of phosphonic derivatives on the metal surface is demonstrated in **Figure 10**, which is shows the obtained AFM images of carbon steel in the presence of diethyl((4-(dimethylamino)phenyl)(phenylamino)methyl)phosphonate and 4-(2-{[ethoxy(hydroxy)phosphonyl](3-nitrophenyl)methyl}hydrazinyl)benzoic acid [1, 16].

The analysis of the achieved AFM images indicates that the addition of phosphonates or phosphonic acids makes a significant modification on the surface morphology of the metal indicating the formation of a protective adsorbed layer of

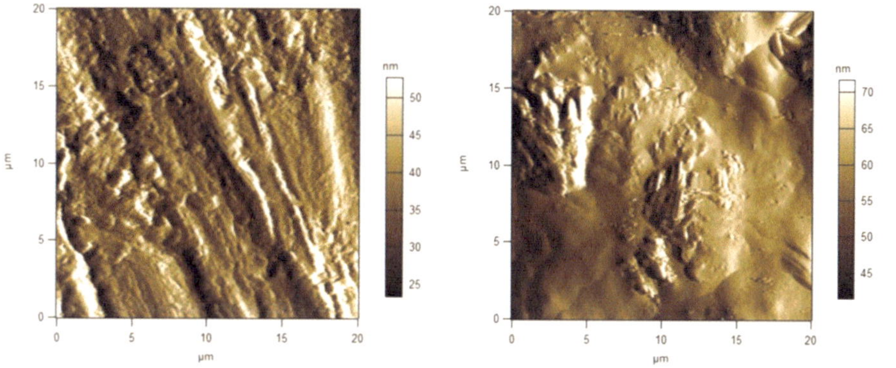

(a) (b)

Figure 10.
AFM images of mild steel in the presence of 10^{-3} mol/L of diethyl ((4-(dimethylamino) phenyl) (phenylamino) methyl) phosphonate (a) and 4-(2-{[ethoxy(hydroxy)phosphonyl](3-nitrophenyl)methyl}hydrazinyl)benzoic acid (b) [1, 16].

phosphonate molecules. This phenomenon can be explicated by the diminution of the measured values of the roughness of metal after the addition of phosphonates or phosphonic acids to the corrosive media.

3.6 Density functional theory (DFT)

Recently, the quantum chemical calculations by applying DFT method are extensively used to correlate the experimental results of corrosion inhibition efficiencies with some quantum descriptors, electronic and structural proprieties of the inhibitive molecule such as the energy of the Highest Occupied Molecular Orbital (E_{HOMO}), the energy of the Lowest Unoccupied Molecular Orbital (E_{LUMO}), the energy gap (ΔE_{gap}), the absolute electronegativity (χ), the hardness (η), the softness (σ), and the dipole moment (μ). Basing on our previously published works, the DFT results indicated that the high corrosion inhibition activity of phosphonates and phosphonic acids is related to the presence of phosphonate or phosphonic acid functional groups in their molecular structure. Also, DFT calculations on phosphonates and phosphonic acids proved that the active sites responsible for the anticorrosion activity of these derivatives are located on the heteroatoms such as P, O, N and S. On the other hand, the DFT study demonstrates that the most negative sites responsible for the electrophilic attacks are located on the oxygen atoms of the phosphonate groups. The **Table 1** summarized the calculated quantum chemical descriptors of two phosphonate derivatives [22].

Quantum descriptors	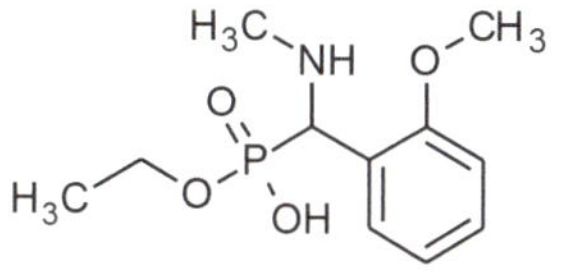	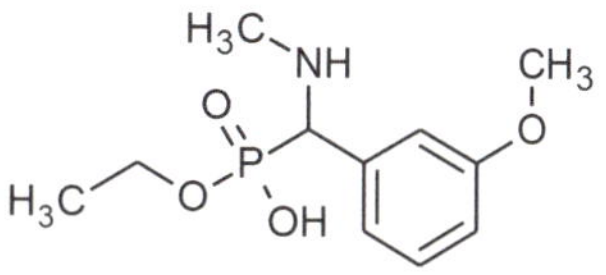
E_{HOMO} (eV)	−6.00039	5.99494
E_{LUMO} (eV)	−0.27620	0.23402
ΔE_{gap} (eV)	5.91386	5.76092
χ (eV)	3.13829	3.11448

Quantum descriptors	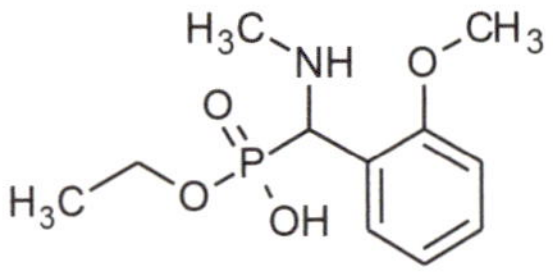	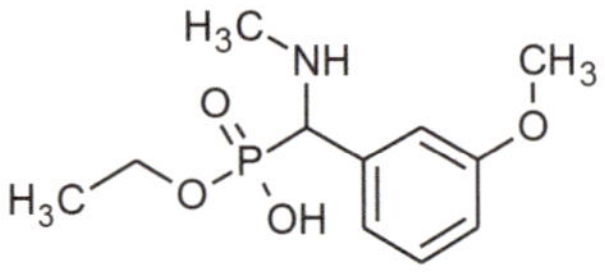
η (eV)	2.86209	2.88046
σ (eV)	0.34939	0.34716
μ (Debye)	2.9461	2.4875

Table 1.
Some calculated quantum chemical descriptors of ethyl hydrogen [(2-methoxyphenyl)(methylamino) methyl] phosphonate and ethyl hydrogen [(3-methoxyphenyl) (methylamino) methyl]phosphonate [22].

3.7 Molecular dynamic simulations (MDS)

Molecular dynamic simulations (MDS) are one of the most effective theoretical methods for elucidates and interprets at the molecular level the mode and configurations of the adsorbed inhibitory molecules on the metal surface. Also, the MDS method is principally used to calculate the adsorption energy of the inhibitive molecules. In this context, MDS are largely applied to analysis the adsorption modes of phosphonates and phosphonic acids on metal surfaces. The adsorption energies of phosphonate derivatives on metal surfaces can be also calculated using MDS.

Concerning the application of MDS to study the mechanism of adsorption of phosphonate molecules on metal surfaces, we can cite the results obtained by Moumeni et al. [23], where she studied the adsorption mode of three phosphonic derivatives on the carbon steel surface (**Figure 11**). Also, the calculated values of the adsorption energy for the investigated inhibitors indicate that adsorption affinity of the substituted phosphonate in *Para* position is higher than these of *Ortho* and *Meta* positions (**Table 2**).

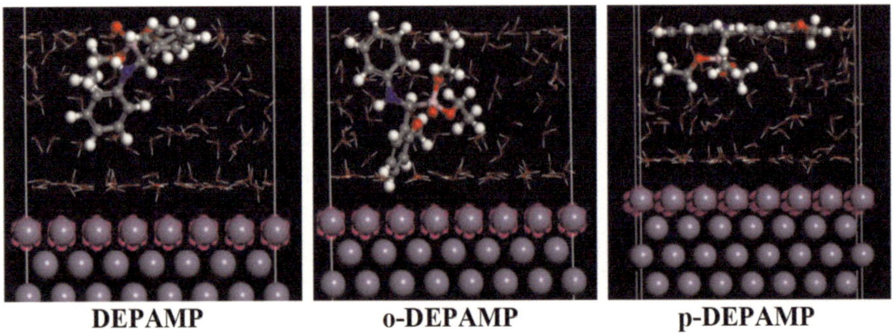

Figure 11.
AFM image of mild steel in the presence of 10^{-3} mol/L of ethyl hydrogen [(2-methoxyphenyl)(methylamino) methyl]phosphonate [23].

Phosphonic inhibitor	DEPAMP	o-DEPAMP	p-DEPAMP
Adsorption energy (kJ/mol)	−6589.8	−6614.9	−6773.9

Table 2.
Calculated adsorption energies of DEPAMP, o-DEPAMP and p-DEPAMP using molecular dynamic simulations [23].

Furthermore, the adsorption energy of phosphonates and phosphonic acids increases with the increase in the number of phosphonate groups in the molecule (O=P—$(OH)_2$ or O=P—$(OR)_2$).

Figure 11 shows the MDS results of Diethyl(phenyl(phenylamino)methyl) phosphonate (DEPAMP), Diethyl(((2methoxyphenyl)amino)(phenyl)methyl) phosphonate (o-DEPAMP) and Diethyl(((4methoxyphenyl)amino)(phenyl)methyl) phosphonate (p-DEPAMP) [23].

4. Conclusion

As a conclusion, both experimental and theoretical techniques prove that phosphonates and phosphonic acids are promising corrosion inhibitors and their high inhibitory activity is directly linked to the presence of phosphonic groups (O=P—$(OH)_2$ or O=P—$(OR)_2$) in their molecular structure. Also, the mechanism of action of this type of inhibitors consists to form an adsorbed layer of the inhibitive molecules on the metal surface. On the other hand, the electrochemical techniques show that phosphonates and phosphonic acids act as mixed type of inhibitor by controlling both anodic and cathodic reactions. Moreover, the microscopic techniques such as SEM and AFM confirm the formation of a protective thin layer of the adsorbed molecules on the metal surface, and this phenomenon is confirmed by the decrease in the values of roughness of metal surface after the addition of inhibitors to aggressive solution. Finely, the theoretical methods such as DFT and MDS demonstrate that the active sites responsible for the adsorption of these molecules are located on the oxygen atoms of the phosphonic functional groups.

Acknowledgements

This research was supported by the General Directorate for Scientific Research and Technological Development (DGRSDT), Algerian Ministry of Scientific Research, Laboratory of Electrochemistry of Molecular Materials and Complex (LEMMC), Ferhat ABBAS University of Sétif.

Conflict of interest

The authors declare no conflict of interest.

Author details

Nadjib Chafai* and Khalissa Benbouguerra
University Ferhat Abbas of Setif, Faculty of Technology, Laboratory of Electrochemistry of Molecular Materials and Complex (LEMMC), Department of Process Engineering, Sétif, Algeria

*Address all correspondence to: n.chafai@univ-setif.dz

References

[1] Chafai N, Chafaa S, Benbouguerra K, Daoud D, Hellal A, Mehri M. Synthesis, characterization and the inhibition activity of a new α-aminophosphonic derivative on the corrosion of XC48 carbon steel in 0.5 M H_2SO_4: Experimental and theoretical studies. Journal of the Taiwan Institute of Chemical Engineers. 2017;**70**:331-344. DOI: 10.1016/j.jtice.2016.10.026

[2] Kerkour R, Moumeni O, Djenane M, Chafai N, Saleh CS. Synergistic corrosion inhibitor of carbon steel by dihydroxy benzyl phosphonic acid-Zn^{2+} system in 0.5 M H_2SO_4: Experimental and theoretical studies. Indian Journal of Chemical Technology. 2022;**29**:402-411. Available from: http://nopr.niscpr.res.in/handle/123456789/60151

[3] Kerkour R, Chafaa S, Maouche N, Moumeni O, Chafai N. Corrosion inhibition of stainless steel N304 by dihydroxy benzyl phosphonic acid in 0.5 M H_2SO_4: Experimental and theoretical studies. Indian Journal of Chemical Technology. 2019;**26**:69-75. DOI: 10.56042/ijct.v26i1.12896

[4] Tabti L, Khelladi RM, Chafai N, Lecointre A, Nonat AM, Charbonnière LJ, et al. Corrosion protection of mild steel by a new Phosphonated pyridines inhibitor system in HCl solution. Advanced Engineering Forum. 2020;**36**:59-75. DOI: 10.4028/www.scientific.net/AEF.36.59

[5] Michaelis A, Kaehne R. Ueber das Verhalten der Jodalkyle gegen die sogen. Phosphorigsäureester oder O-Phosphine. Berichte. 1898;**31**:1048-1055. DOI: 10.1002/cber.189803101190

[6] Arbuzov AE. Reactions of alkyl halides with phosphites. Russian Journal of Physical Chemical Society. 1906;**38**:687

[7] Kabachnik MI, Medved TY. Новый метод синтеза сс-аминофосфиновых кислот. Doklady Akademii Nauk SSSR. 1952;**83**:689-692

[8] Fields EK. The synthesis of esters of substituted amino phosphonic acids. Journal of the American Chemical Society. 1952;**74**:1528-1531. DOI: 10.1021/ja01126a054

[9] Pudovik AN, Zametaeva GA. New method for the synthesis of phosphonic and phosphinic esters and their thio analogs communication 13. Addition of O,O-diethyl hydrogen phosphorothidite to ketones and aldehydes. Russian Chemical Bulletin. 1952;**1**:825-830. DOI: 10.1007/BF01198872

[10] Abramov VS. Reaction of dialkyl phosphites with aldehydes and ketones. A new method of preparation of esters of hydroxyalkanephosphonic acids. Zhurnal Obshchei Khimii. 1952;**22**: 647-652

[11] Moedritzer K, Irani RR. The direct synthesis of α-Aminomethylphosphonic acids. Mannich-type reactions with Orthophosphorous acid. The Journal of Organic Chemistry. 1966;**31**:1603-1607. DOI: 10.1021/jo01343a067

[12] Tlidjane H, Chafai N, Chafaa S, Bensouici C, Benbouguerra K. New thiophene-derived α-aminophosphonic acids: Synthesis under microwave irradiations, antioxidant and antifungal activities, DFT investigations and SARS-CoV-2 main protease inhibition. Journal of Molecular Structure. 2022;**1250**: 131853. DOI: 10.1016/j.molstruc.2021.131853

[13] Kerkour R, Chafai N, Moumeni O, Chafaa S. Novel α-aminophosphonate derivates synthesis, theoretical

calculation, molecular docking, and in silico prediction of potential inhibition of SARS-CoV-2. Journal of Molecular Structure. 2023;**1272**:134196. DOI: 10.1016/j.molstruc.2022.134196

[14] Elkolli M, Chafai N, Chafaa S, Kadi I, Bensouici C, Hellal A. New phosphinic and phosphonic acids: Synthesis, antidiabetic, anti-Alzheimer, antioxidant activity, DFT study and SARS-CoV-2 inhibition. Journal of Molecular Structure. 2022;**1268**:133701. DOI: 10.1016/j.molstruc.2022.133701

[15] Zaout S, Chafaa S, Hellal A, Boukhemis O, Khattabi L, Merazig H, et al. Hydroxyphenylamine phosphonate derivatives: Synthesis, X-ray crystallographic analysis, and evaluation of theirs anti-Alzheimer effects and antioxidant activities. Journal of Molecular Structure. 2021;**1225**:129121. DOI: 10.1016/j.molstruc.2020.129121

[16] Benbouguerra K, Chafaa S, Chafai N, Mehri M, Moumeni O, Hellal A. Synthesis, spectroscopic characterization and a comparative study of the corrosion inhibitive efficiency of an α-aminophosphonate and Schiff base derivatives: Experimental and theoretical investigations. Journal of Molecular Structure. 2018;**1157**:165-176. DOI: 10.1016/j.molstruc.2017.12.049

[17] Cao C. On electrochemical techniques for interface inhibitor research. Corrosion Science. 1996;**38**: 2073-2082. DOI: 10.1016/S0010-938X(96)00034-0

[18] Ouksel L, Bourzami R, Chafaa S, Chafai N. Solvent and catalyst-free synthesis, corrosion protection, thermodynamic, MDS and DFT calculation of two environmentally friendly inhibitors: Bis-phosphonic acids. Journal of Molecular Structure. 2020;**1222**:128813. DOI: 10.1016/j.molstruc.2020.128813

[19] Tan KW, Kassim MJ. A correlation study on the phenolic profiles and corrosion inhibition properties of mangrove tannins (Rhizophora apiculata) as affected by extraction solvents. Corrosion Science. 2011;**53**: 569-574. DOI: 10.1016/j.corsci.2010.09.065

[20] Labjar N, Lebrini M, Bentiss F, Chihib NE, El Hajjaji S, Jama C. Corrosion inhibition of carbon steel and antibacterial properties of aminotris (methylnephosnic) acid. Materials Chemistry and Physics. 2010;**119**: 330-336. DOI: 10.1016/j.matchemphys.2009.09.006

[21] Bentiss F, Lebrini M, Lagrenée M. Thermodynamic characterization of metal dissolution and inhibitor adsorption processes in mild steel/2,5-bis (n-thienyl)-1,3,4- thiadiazoles/ hydrochloric acid system. Corrosion Science. 2005;**47**:2915-2931. DOI: 10.1016/j.corsci.2005.05.034

[22] Djenane M, Chafaa S, Chafai N, Kerkour R, Hellal A. Synthesis, spectral properties and corrosion inhibition efficiency of new ethylhydrogen [(methoxyphenyl)(methylamino) methyl]phosphonate derivatives: Experimental and theoretical investigation. Journal of Molecular Structure. 2019;**1175**:398-413. DOI: 10.1016/j.molstruc.2018.07.087

[23] Moumeni O, Chafaa S, Kerkour R, Benbouguerra K, Chafai N. Synthesis, structural and anticorrosion properties of diethyl(phenylamino)methyl phosphonate derivatives: Experimental and theoretical study. Journal of Molecular Structure. 2020;**1206**:127693. DOI: 10.1016/j.molstruc.2020.127693

Chapter 3

Controlling Corrosion Using Non-Toxic Corrosion Inhibitors

Malak Rehioui

Abstract

Corrosion of metals and its alloys destroys our properties, our environment, and our lives. Thus, corrosion control includes a range of developed treatments that take into account material properties, environmental characteristics, and process cost. Typical corrosion inhibitors are known for their excellent efficiency and show great promise. However, they become less used because they cause serious toxicity issues on the environment and affect human and animal health. In recent years, research has intensified on the development of green chemistry technologies, which offer new methods of synthesis and extraction of various non-toxic materials (plant extracts, oils, amino acids, rare earths, etc.), which are highly effective, environmentally acceptable, economical and easily available inhibitors. This chapter deals with a description of corrosion inhibitors with a particular emphasis given to the discussion on the different characteristic features of the green corrosion inhibitors reported in the literature as a comparative view of toxic inhibitors.

Keywords: corrosion control, metals, green inhibitors, toxicity, environment

1. Introduction

Corrosion is an unavoidable natural phenomenon. It is the destruction or deterioration over time of metals and alloys and the alteration of their composition and their physical properties caused by a reaction with the surrounding environment [1]. The tendency of metals to corrode depends on several factors; involving the reactivity of the metal, the presence of impurities, the presence of air, moisture, gases such as sulfur dioxide and carbon dioxide, and the presence of corrosive electrolytes [1, 2]. In addition, corrosion process is defined as the spontaneous tendency of the material to revert back to its original state as found in nature which is more thermodynamically stable form. For this reason, corrosion is also called the reverse of extractive metallurgy [3]. Corrosion processes are a constant and continuous problem that develop rapidly and cause significant damage to society as they deteriorate structural innovations and engineering materials, etc. Corrosion is expensive due to the loss of materials or their properties [4, 5]. Meanwhile, the environmental consequences of corrosion are both severe and complex. They generally extend far beyond the immediate issue of resource depletion. In some cases may be toxic and cause injury [5]. Therefore, corrosion is a subject of great importance due to its economic and safety concerns. This is why metals and alloys require protection in various process industries. Nevertheless, by implementation of corrosion prevention strategies in an appropriate manner, the metal degradation cost can be reduced. At the present time, corrosion control comprises an array of available treatments and approaches that have been developed, such as the isolation of the structure from aggressive media using coatings or the compensation for the loss of electrons from the corroded structure (e.g. cathodic protection by impressed current or by using active sacrificial anodes), or the use of corrosion inhibitors, etc. [6]. Among them, corrosion inhibitors have proved themselves effective in protecting metals against corrosion with obvious advantages regarding availability, strong adaptability, economic efficiency, and high protection efficiency. A corrosion inhibitor is defined as a chemical substance that is added in small concentration to the corrosion medium, which leads to a decrease in the corrosion rate of the metal. It fights against corrosion without directly treating the metal, but intervenes through the medium. Its effectiveness depends on its ability to react with the surface of the metal to form a protective film, thereby reducing or providing protection against corrosion [7–9].

However, the choice of corrosion inhibitors must also be consistent with non-toxicity criteria, since most traditional corrosion inhibitors have been considered highly toxic to living systems and have negative environmental impacts [9, 10]. Hazards arising from the toxicity of regular inhibitors have created the need to develop and explore highly effective and non-toxic inhibitors called "green inhibitors." It is based on natural products or plant extracts, oral medicines, food spices, rare earths, etc. The concept of green corrosion inhibitors has gained popularity as environmental awareness has increased dramatically. In this respect, this chapter is intended to present at first regular corrosion inhibitors and show a detailed description of their toxicity. Then, it provides an in-depth view at the contemporary studies on non-toxic natural product inhibitors with their sources, as well as a brief description of their mechanisms for preventing corrosion.

2. Frequently utilized corrosion inhibitors

Corrosion inhibitors form a defensive barrier of one or several molecular layers against corrosive agent attack. The frequently used inhibitors can be divided into two main types, inorganic substances, and organic substances.

Indeed, inorganic inhibitors are those in which the active substance is an inorganic compound [10]. Their metallic atoms are enclosed in the film to improve corrosion resistance [11]. Many inorganic inhibitors are known for their excellent efficacy. In particular, chromium (Cr^{6+}) or chromium compounds and chromium salts are among the passivating inhibitors par excellence, but they are environmentally unacceptable and their severe toxicity greatly reduces their use. As has been indicated, strontium chromate, zinc chromate, chrome phosphate, etc., are heavy-metal-based and highly carcinogenic. Small amounts of chromic acid or potassium dichromate can cause kidney failure, liver damage, DNA damage, and blood cell disorders. Chromate mists entering the lungs may eventually lead to lung cancer [12, 13]. Other inhibiting molecules in the form of salts have made it possible to obtain very good yields in terms of metal protection and thus corrosion prevention, such as sodium tungstate (Na_2WO_4), vandates ($NaVO_3$), nitrites ($NaNO_2$), and silicates ($Na_2Si_2O_5$). However, these compounds today are highly toxic, leading to serious consequences for the environment and human beings. They can cause temporary or permanent damage to the nervous system, and disrupt the biochemical process and the enzymatic system of our organism [9]. Molybdates (MoO_3^-) and phosphates ($H_2PO_3^-$) also provide passivation protection to metallic surfaces by incorporating them into the oxide layer. Borates and arsenates are also known for their promising inhibitory activity against metal corrosion in various aggressive aqueous media. Apart from that, they have also proven to be intolerant because of the threat they pose to nature and social health in the long run [14, 15]. Pyrrole and derivatives exhibit good protection against metals corrosion, especially in acidic media. These inhibitors are also useful application in the formulation of primers and anticorrosive coatings, but the major disadvantage associated with them is their toxicity and as such their use has come under severe criticism [15].

On the other hand, the frequently synthetic organic molecules used as corrosion inhibitors in industrial environments include aliphatic or aromatic thioureas, amines, amides, pyrazoles, pyrimidines, acetylenic alcohols, aldehydes, benzylidenes, carbazones, azoles, Schiff's base, benzonitriles, dimeric and trimeric acids, etc. Indeed, most of these inhibitors are containing heteroatoms such as nitrogen, sulfur, and oxygen with lone pair of electrons and should have aromatic systems. These compounds can act on the metal surface by means of adsorption and there detracting of the active metallic surface area, leaving inactive sites on the surface exposed to corrosive media. The inhibition efficiency of these compounds is also related to its functional groups, steric effects, and π-orbital character of donating electrons [15]. Even though these organic compounds exhibit high inhibition efficiencies against the corrosion of many types of metals, they are toxic and non-environmentally friendly and their use causes toxic harm to humans, animals, and nature. In addition, the time of exposure is also a factor that can have a strong influence on the toxicity and which can increase the harmful effect of these molecules [13].

So, the environmental and health risks associated with the use of these inhibitors have prompted us to find or use non-toxic or green corrosion inhibitors that would offer maximum protection to metal structures but have minimal impact on human and nature. Because the choice of an effective corrosion inhibitor must not only be cost-effective, stable, compatible with the corrosive medium, and produce the desired effect at small concentrations; but it must also be compatible with the current standards for non-toxicity, biodegradability, bioaccumulation, and environmental protection.

3. Non-toxic corrosion inhibitors

The toxicity of inhibitors and the increase in environmental pollution have led to the enactment of strict international laws for the use of ecological inhibitors and to the demand for a green and ecological approach that deals with the principles of "green chemistry". These principles refer to efforts toward establishing a comprehensive approach to chemical risk management. This concept is based on the ideas of sustainability, reducing environmental consequences, and preserving natural resources for the following centuries [16]. The requirements for a chemical to be approved as a green corrosion inhibitor have been explicitly set out by legislative bodies essentially the Paris Commission (PARCOM) and the Registration, Evaluation, Authorization, and Restriction of Chemicals (REACH), which are non-bio-accumulative, biodegradable, and zero or very minimal marine toxicity level [17]. **Figure 1** illustrates the principles by which green corrosion inhibitors work.

Figure 1.
The principles of green chemistry on which green corrosion inhibitors act.

The major categories of these kinds of inhibitors include plant extracts, oils, rare earths, and amino acids have been discussed briefly below. Other green inhibitors such as ionic liquids, organic polymers, and surfactants [18–20] are not emphasized in this chapter.

Green corrosion inhibitors are drawing extraordinary interest in the corrosion field and can be used in many industrial applications [11] to promote green chemistry and sustainability. In this perspective, one of the areas in which green chemistry is associated with green inhibitors, through the reduction of ecological impact and wastes, is related to safeguarding metals. **Figure 2** shows the importance of green corrosion inhibitors in large industrial applications.

3.1 Rare earth metal compounds

The use of rare earth metal compounds as corrosion inhibitors traced back to 1984 when Hinton et al. [21] published the first paper on the use of cerium chloride salts as corrosion inhibitors, after that a lot of research papers were published and showed that rare earth metal compounds can be used as good alternatives of non-toxic corrosion inhibitors. In 1992, Hinton et al. [22] published a review paper highlighting the use of some rare earth salts as green corrosion inhibitors for a wide range of metals. Rare earth compounds act by producing an oxide film at the cathodic sites of metal substrates that avoid the supply of oxygen or electrons to the reduction reaction, thus minimizing the rate of corrosion. The majority of rare earths metals have zero toxicity [10]. Thus, recent research turns around utilizing rare earth metals as green alternative for toxic inhibitors, especially chromium species.

For instance, Somers AE et al. [23] evaluated four rare earth 3-(4-methylbenzoyl) propanoate (mbp) compounds (RE = La, Ce, Nd, and Y) as corrosion inhibitors for

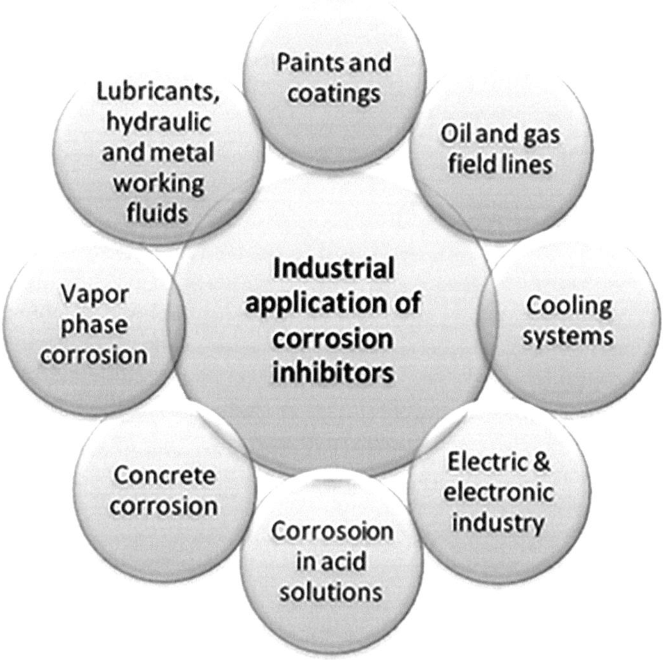

Figure 2.
Application of green corrosion inhibitors in various industrial sectors.

mild steel in 0.01 M NaCl. Results showed that all the compounds can reduce corrosion after 30 min immersion. Surface analysis showed the presence of a film containing inhibitor components.

In another investigation of Manh TD et al. [24] the rare-earth organic compound Gadolinium 4-hydroxycinnamate ($Gd(4OHCin)_3$) was shown to be an effective corrosion inhibitor for mild steel in naturally-aerated 0.1 M chloride solution, not only of general corrosion but also of pitting corrosion. The inhibition efficiency is more important when the concentration of inhibitor increases, reaching values up to 94% at a concentration of 0.93 mM. The results also demonstrated that $Gd(4OHCin)_3$ behaves as good mixed corrosion inhibitor with predominant anodic activity.

Peng Y et al. [25] studied two novel rare-earth (RE) 3-(4-methylbenzoyl)-propanoate (mbp) complexes ($RE(mbp)_3$; RE = La, Y) as corrosion inhibitors for AS1020 mild steel in 0.01 M NaCl solutions. Results disclosed a high corrosion inhibition performance of $Y(mbp)_3$ which is attributed to the build-up of a protective surface film with a high level of corrosion resistance, particularly after 24 hours.

Zhao D et al. [26] focused on the use of the salt of rare earth cerium as corrosion inhibitor of aluminum. Results, revealed that the good corrosion resistance of cerium-based passive coating was obtained when the compositions were as follows: $CeCl_3{\cdot}7H_2O$, 0.05 mol/L; H_2O_2, 30 mL/L; current density, 1.1 mA/cm_2; temperature, 40°C; time, 9 min. Surface analysis showed that the cerium conversion coatings formed on the surface of aluminum alloy were related to cerium hydroxide/hydrated oxide depositions.

Porcayo-Calderon J et al. [27] evaluated the corrosion inhibition effect of rare earth chlorides on API X70 steel in a 3.5% NaCl solution by electrochemical techniques. The results showed that it is a mixed-style inhibitor with an inhibition efficiency greater than 90%, at a concentration of 0.001 M. Its protective action is due to the reduction of the oxygen reduction rate because of the blocking effect of the cathodic sites and to the reduction of the metallic dissolution rate due to the formation of a protective layer on metal surface.

3.2 Plant extracts and oils

The use of natural products to inhibit corrosion may date back to the 1960s when tannins and their derivatives were employed to protect steel, iron tools, and pipelines [28, 29]. Now there are many studies, articles, reviews and books focused on the development of metal corrosion inhibitors, that guarantee high efficiency up to 99%, based on plant extracts and oils rich in active molecules, obtained from different parts of plant-like leaves, fruits, bark, peels, flowers, roots, seeds, stems, and even whole plant extracts. Plants are eco-friendly climate as they prepare their food through the photosynthesis cycle by taking carbon dioxide and releasing oxygen. In addition, the plant extract is also environmentally friendly when used as an inhibitor as these are likewise biodegradable.

The inhibition efficiency of these green inhibitors is due to the presence of phytochemicals [30]. Most phytochemicals contain polar functional groups such as amide ($-CONH_2$), hydroxyl ($-OH$), ester ($-COOC_2H_5$), carboxylic acid ($-COOH$), and amino ($-NH_2$) which aid in their absorption on metal surface [31]. Phytochemical type and content vary based on the choice of plant components for extraction. Some of the most common phytochemicals that have a corrosion-inhibiting effect are flavonoids, glycosides, alkaloids, saponins, phytosterols, tannins, anthraquinones, phenolic compounds, triterpenes, and fluoptanins. Among them, we quote some published works:

The flavonoid extract from *Erigeron floribundus* was studied as green inhibitor for mild steel corrosion in 2 M HCl solution using gasometric method. The study revealed that the inhibition efficiency increased with increase in concentration of the inhibitor. The adsorption mechanism was spontaneous and occurred according to Langmuir adsorption isotherms with also physical adsorption [32].

Fouda AS et al. [33] have studied the extracts of henna (*Lawsonia Inermis*) for corrosion inhibition of carbon steel in 1 M HCl solution applying weight loss and electrochemical measurements. Results showed that the inhibition efficiency increases with increasing inhibitor concentration and reached 83.1% at 300 ppm, however, it decreases with increasing temperature. Surface analysis has been carried out using energy-dispersive X-ray and scanning electron microscopy.

Rehioui et al. [34] explored the anticorrosion behavior of the *Opuntia dillenii* seed oil incorporated in a formulation labeled FOD as an ecofriendly corrosion inhibitor to protect iron in acid rain. Corrosion inhibition effect of FOD was studied by gravimetric methods, electrochemical measurements, and scanning electron microscopy coupled with elemental analysis (SEM/EDX). Obtained results revealed that FOD acted as a good mixed corrosion inhibitor with predominant anodic activity. Inhibition efficiency was found to vary with concentration and period of immersion, reaching values up to 99% at the concentration of about 1000 ppm. The adsorption study showed that it followed Langmuir adsorption isotherm with both chemisorption and physisorption mechanism.

Torres-Acosta AA [35] has investigated *Opuntia-Ficus-Indica* (Nopal) mucilage as a steel corrosion inhibitor in alkaline media. Results showed good corrosion-inhibiting effect of *Opuntia-Ficus-Indica* (Nopal) mucilage. The addition of Nopal led to the formation of a denser and more packed oxide/hydroxide surface layer on the steel surface that decreased corrosion activity. This oxide/hydroxide layer growth was confirmed from microscopic evaluation of the metal surface.

The performances of the extract obtained from *Rosmarinus officinalis* (*RO*) on the corrosion inhibition of XC48 steel in 1 M HCl at different temperatures were carried out through mass loss, electrochemical measurements, surface analysis, and quantum chemical calculations. Results showed that *RO* extract is a mixed-type inhibitor. The inhibition efficiency increased at greater concentration of the inhibitor and decreases with the rise of the temperature. The adsorption mechanism is physisorption that is adequately described by the Langmuir equilibrium model. The retrieved outcomes are confirmed by surface observations, which reveal that the adsorbed inhibitor molecules completely hinder the HCl attacks at the steel grain boundaries [36].

Chellouli et al. [37] determined the inhibitive effect of a green formulation based on the seed oil of *Nigella Sativa* L. against iron corrosion in acid rain solution by applying gravimetric methods, electrochemical measurements, and surface analysis. Results demonstrated that the formulation acts as a good mixed-type inhibitor. The metal dissolution rate decreased with increasing inhibitor concentration and immersion time. A maximum inhibition efficiency of around 99% is achieved for a concentration of 2500 ppm. The surface analysis confirmed a good protective action of the inhibitor by the formation of a film on the surface of the iron in an environment simulated with acid rain.

The performance of other recently developed plant extracts and oils as green corrosion inhibitors of different metals and alloys in various aggressive media is listed in **Table 1**.

Inhibitor (concentration)	Metal/alloy	Test condition	Test techniques	Maximum efficiency (%)	Reference
Viscum Album extract (300 ppm)	Carbon steel	1 M HCl	Weight loss Electrochemical measurements Atomic force microscopy Attenuated total reflection infrared X-ray spectroscopy	96.30	[38]
Apricot almond oil (0.5 g/L)	Steel	1 M HCl	Weight loss Electrochemical measurements	83.49	[39]
Jatropha Curcas Seed oil (250 ppm)	Iron	Acid rain	Electrochemical measurements Scanning electronic microscopy coupled with energy-dispersive	97	[40]
Galactomannan extract from *Ceratonia siliqua* L (1 g/L)	Iron	1 M HCl	Weight loss Electrochemical measurements UV visible Computational chemistry calculations Scanning electronic microscopy coupled with energy-dispersive	87.72	[41]

Inhibitor (concentration)	Metal/alloy	Test condition	Test techniques	Maximum efficiency (%)	Reference
Passion fruit seed oil microemulsion	P110 Carbon steel	CO_2^- saturated brine	Weight loss Electrochemical measurements Scanning electron microscopy Contact angle measurement	99	[42]
Matricaria recutita chamomile extract	S235JR steel	0.5 M NaCl	Electrochemical measurements Computational chemistry studies Scanning electron microscope coupled with energy-dispersive X-ray Fourier transform infrared spectroscopy	98.90	[43]
Carica papaya peel extracts	Dual phase steel	3.5 wt % NaCl	Electrochemical measurements	98.67	[44]
Piper longum extract (400 mg/L)	Al-1060 aluminum	1 M NaOH	Weight loss Electrochemical measurements	94	[45]
Bee pollen extract (7 g/L)	Copper	1 M HCl	Weight loss Electrochemical measurements Fourier-transform infrared spectroscopy X-ray photoelectron spectroscopy Atomic force microscopy Scanning electron microscopy	94.50	[46]
Arbutus unedo L. leaves extract (0.5 g/L)	Mild steel	1 M HCl	Electrochemical measurements Scanning Electron Microscopy Diffuse reflectance infrared Fourier transform Computational chemistry studies	91.72	[47]
Leonurus japonicus Houtt. extract (400 mg/L)	Copper	0.5 mol/L H_2SO_4	Electrochemical measurements Fourier transform infrared spectroscopy X-ray photoelectron spectroscopy Computational chemistry calculations	90	[48]
Roselle (*Hibiscus sabdariffa*) leaf extract	Cu-Zn alloy	1 M HNO_3	Weight loss Electrochemical measurements Scanning electron microscopy Energy-dispersive X-ray	94.89	[49]

Inhibitor (concentration)	Metal/alloy	Test condition	Test techniques	Maximum efficiency (%)	Reference
Thymus vulgaris extract (800 ppm)	Bronze	Acid rain	Gravimetric and electrochemical measurements	90	[50]
Cymbopogon schoenanthus aerial extract (250 ppm)	Aluminum brass	Acid cleaning solutions	Weight loss Electrochemical measurements Scanning electronic microscopy Fourier transform infrared spectroscopy	97	[51]
Ocimum basilicum essential oil	C38 Steel	0.5 M H_2SO_4	Electrochemical measurements Computational chemistry calculations	88.10	[52]
Citrus reticulata peel's essential oil (900 ppm)	Mild steel	1 M HCl	Weight loss Electrochemical measurements Scanning electronic microscopy coupled with energy-dispersive X-ray spectrometry Computational chemistry calculation	90.30	[53]
Apigenin isolated from *Hypericum perforatum* (30 mg/L)	Brass	1 M HNO_3	Weight loss Electrochemical measurements Scanning electron microscope Atomic force microscope X-ray photoelectron spectroscopy Raman spectroscopy measurements	90	[54]
Cabbage extract	X70 Steel	1 M HCl	Electrochemical measurements Scanning electron microscope Atomic force microscope X-ray photoelectron spectroscopy Computational chemistry calculations	95.87	[55]
Orange peel extracts (0.03%)	AZ91D Magnesium alloy	0.05 wt.% NaCl	Electrochemical measurements Scanning electron microscope Atomic force microscopy X-ray diffractometer Density functional theory Fourier transform infrared spectroscopy	85.70	[56]

Inhibitor (concentration)	Metal/alloy	Test condition	Test techniques	Maximum efficiency (%)	Reference
Sapium ellipticum leaf extract (1.5 g/L)	AA3003 Aluminum alloy	1 M HCl	Electrochemical measurements Response surface methodology	96.73	[57]
Essential oil from aerial parts of *Artemisia herba-alba* (1 g/L)	Stainless Steel	1 M H_3PO_4	Electrochemical measurements Scanning electron microscope coupled with energy-dispersive X-ray	88	[58]
Eydrosol extract of *Thymbra capitata* (L.) Cav. (Lamiaceae)	Brass	3% NaCl	Electrochemical measurements Scanning electron microscope coupled with energy-dispersive X-ray Computational chemistry calculations	93.04	[59]

Table 1.
Some plant extracts as corrosion inhibitors of different metals and alloys.

3.3 Amino acids

Amino acids are considered as green corrosion inhibitors because they are non-toxic, biodegradable, inexpensive, soluble in aqueous media, and easy to produce purities higher than 99%. Amino acids are organic compounds that contain at least one carboxyl group (–COOH) and one amino group (–NH_2) bonded to the same carbon atom (α- or 2-carbon) [60]. The presence of heteroatoms and conjugated π-electrons system have made amino acids a significant class of green corrosion inhibitors thanks to their environmental aspect. It has been demonstrated by various authors that certain amino acids have been shown to be good and reliable corrosion inhibitors for many metals in various aggressive environments, which has led to a growing interest in these compounds as alternatives to conventional corrosion inhibitors, which are often toxic, as mentioned in the previous section. However, from then, the number of studies dealing with amino acids as corrosion inhibitors increased rapidly. In other respects, amino acids are used in food and feed technology and as intermediates for the chemical industry (e.g. for pharmaceutical and cosmetic applications) [61].

Among the amino acids, El-Sayed NH [62] signaled the corrosion inhibition of carbon steel in stagnant naturally aerated chloride solutions by certain amino acids including glycine, valine, leucine, cysteine, methionine, histidine, threonine, phenylalanine, lysine, proline, aspartic acid, arginine, and glutamic acid using electrochemical techniques. Results showed that all of the amino acids acted as mixed-style inhibitors while cysteine, phenylalanine, arginine, and histidine showed remarkably high corrosion inhibition efficiency at a concentration of 10 mM/dm^3.

Amin AM et al. [63] studied corrosion inhibition of copper in O2-saturated 0.5 M H_2SO_4 solutions by four selected amino acids glycine, alanine, valine, and tyrosine, using electrochemical measurements at 30°C. The inhibition efficiencies of almost 98 and 91% were obtained with 50 mM tyrosine and glycine, respectively. On the other hand, alanine and valine reached only about 75%.

Srivastava V et al. [64] studied the effect of three novel amino acids 2-(3-(carboxymethyl)-1H-imidazol-3-ium-1-yl)acetate (AIZ-1), 2-(3-(1-carboxyethyl)-1H-imidazol-3-ium-1-yl)propanoate (AIZ-2), and 2-(3-(1-carboxy-2-phenylethyl)-1H-imidazol-3-ium-1-yl)-3-phenylpropanoate (AIZ-3) on the corrosion of mild steel by electrochemical methods, surface analysis, and theoretical investigations. Among the studied inhibitors, AIZ-3 showed the maximum inhibition efficiency (IE) of 96.08% at a concentration of 0.55 mM (200 ppm).

Zeino et al. [65] investigated polyaspartic acid (PASP) for corrosion inhibitory effect on mild steel in a 3% NaCl solution. PASP alone showed a moderate inhibition efficiency of 61% at 2 g/L, but when zinc ion was added to PASP, the inhibition efficiency rise to 97% at a reduced PASP concentration of 0.5 g/L.

Amin MA et al. [66] have used glycine derivative to prevent corrosion of mild steel corrosion in 4 M H2SO4 solutions at different temperatures (278–338 K) [37] using electrochemical methods. The inhibition efficiency increased with an increase in inhibitor concentration and decreased with temperature, suggesting the occurrence of physical adsorption.

Zhang DQ et al. [67] investigated the corrosion inhibition of three amino acid compounds namely serine, threonine, and glutamic acid on copper in aerated 0.5 M HCl by electrochemical method, reflected FT-infrared spectroscopy, and quantum chemical calculations.

Some other recently reported amino acid-based as green corrosion inhibitors of variety of metals and alloys are depicted in **Table 2**.

Inhibitor (concentration)	Metal/alloy	Test condition	Test techniques	Maximum efficiency (%)	Reference
Tricine [*N-(Tri(hydroxymethyl)methyl)glycine*] (10 mM)	Zinc	0.5 M NaCl	Electrochemical measurements	90	[68]
L-methionine	AISI309S stainless steel	1 M H_2SO_4	Electrochemical measurements Scanning electron microscopy Atomic force microscopy Adsorption isotherms X-ray photoelectron spectroscopy Contact angle measurement	97	[69]
Lysine (1 g/L)	Low alloy carbon steel	NaCl (1 M H_2SO_4 + 10^{-3} M Cl^-)	Electrochemical measurements	78.88	[70]
Tetra-n-butyl ammonium methioninate	Mild steel	1 M HCl	Electrochemical measurements Scanning electronic microscopy coupled with energy-dispersive X-ray Computational chemistry studies	95.10	[71]

Inhibitor (concentration)	Metal/alloy	Test condition	Test techniques	Maximum efficiency (%)	Reference
Glutamic acid-Zn^{2+} (200–25 ppm)	Carbon steel	Sea water	Weight loss Electrochemical measurements Fourier-transform infrared spectroscopy Scanning electronic microscopy	87	[72]
L-Arginine-Zn^{2+} (250–25 ppm)	Carbon steel	3.5% NaCl	Weight loss Electrochemical measurements Fourier-transform infrared spectroscopy Scanning electronic microscopy Atomic force microscopy Cyclic Voltammetry	91	[73]
L-cysteine (30 mmol/L)	AA5052 aluminum alloy	4 M NaOH	Weight loss Electrochemical measurements Computational chemistry studies	Noticeable efficiency	[74]
L-tryptophan (5.10–2 M)	Mild steel	1 M HCl	Weight loss Electrochemical measurements Computational chemistry studies	92.70	[75]
L-aspartic acid-Zn^{2+}	Carbon steel	Aqueous media	Electrochemical measurements Fourier transform infrared spectroscopy X-ray photoelectron spectroscopy Computational chemistry calculations	90	[76]
Glutathione (0.75 mM)	6061 Al-SiC(p) composite	0.5 M HCl	Weight loss Electrochemical measurements Scanning electron microscopy Energy-dispersive X-ray	80	[77]

Inhibitor (concentration)	Metal/alloy	Test condition	Test techniques	Maximum efficiency (%)	Reference
L-alanine-Zn^{2+} (250–5 ppm)	Carbon Steel	Aqueous medium	Gravimetric and electrochemical measurements Scanning electron microscopy Energy dispersive analysis of X-rays Fourier transform infrared spectroscopy Atomic force microscopy	83	[78]
5-((benzylthio) methyl)- 3-phenyl-2-thioxoimidazolidin-4-one (BPT)	N80 carbon steel	CO_2^- saturated formation water	Electrochemical measurements Scanning electronic microscopy X-ray photoelectron spectroscopy Computational chemistry calculations	99.44	[79]
2-amino-4-methylpentanoic acid (LCN)	Carbon steel	1 M HCl	Electrochemical measurements Weight loss Optical microscopy analysis Computational chemistry calculations	87.46	[80]
Poly(vinyl alcohol cysteine) (0.6 wt%)	Mild steel	1 M HCl	Weight loss Electrochemical measurements Fourier transform infrared spectroscopy UV-visible Scanning electron microscopy Energy dispersive analysis of X-rays	94	[81]
DL-phenylalanine-Zn^{2+} (150–25 ppm)	Carbon steel	Well Water	Weight loss Electrochemical measurements Scanning electron microscopy Energy dispersive analysis of X-rays	90	[82]

Inhibitor (concentration)	Metal/alloy	Test condition	Test techniques	Maximum efficiency (%)	Reference
Protein and its amino acids, isolated from tofu pulp (80 ppm)	Carbon steel	Brackish water media	Electrochemical measurements	92	[83]
l-histidine based ionic liquid (LHIL) (2 mM)	Mild steel	1 M HCl	Electrochemical measurements Scanning electron microscopy Energy dispersive spectroscopy Laser scanning confocal microscope Computational chemistry calculations	98.80	[84]

Table 2.
Some amino acid-based corrosion inhibitors.

4. Mechanism of action of green corrosion inhibitors

During corrosion, metal ions migrate into the solution in the active regions (anodic site) and transfer electrons from the metal to the acceptor at less active regions (the cathode); the cathodic process requires the presence of an electron acceptor functioning as oxygen, oxidizing agents or hydrogen ions. Green corrosion inhibitors have adsorbing properties and are known as site blocking elements [85]. They can minimize the corrosion rate through adsorption of active species onto the metal/alloy surface when added to many industrial systems by:

- The change of the rate of anodic and/or cathodic reactions;
- The influence of the diffusion rate of aggressive ions in interaction with metal structures;
- The increase of the electrical resistance of the metal surface by forming a film on it.

In fact, many researchers have postulated several theories to explain the mode of action of green corrosion inhibitors. For example, the active constituent derived from natural inhibitors varies from one plant species to another. The best sources of green inhibitors are natural products because they contain polar compounds with multiple "heteroatoms" similar to organic inhibitors. These heteroatoms present in the plant extracts act as an active center and adsorb on metallic surface by creating a film that denies access to corrosive agent. Non-polar compounds with aromatic rings, aliphatic chains, heterocyclic rings, and functional moieties are abundant in plant extracts. These compounds can be effectively adsorbed on the mineral surface and thus protect it from corrosion without harming the environment like inorganic compounds [86].

There are several methods to identify the inhibitory mechanism of green corrosion inhibitors. Electrochemical techniques such as electrochemical impedance spectroscopy and potentiodynamic polarization analysis have been successfully implemented and provide valuable information on the corrosion rate and the mechanism of corrosion protection. These methods are briefly described in this section.

Potentiodynamic polarization is an electrochemical method used to determine green corrosion inhibitors performance, instantaneous corrosion rates and to elucidate the corrosion prevention mechanism. This method relies on changing the current or potential across a sample under study and recording the corresponding potential or current change. This can be facilitated using either a direct current source or an alternating current source. In most studies, a conventional three-electrode cell is used for the measurement, consisting of a counter electrode (Pt or graphite), a reference electrode (calomel or Ag/AgCl), and a working elecrode (metal substrate) immersed in the test solution [87]. The reference electrode measures and controls the system's voltage (V) and the counter electrode measures and controls current (I). The open circuit potential (E_{ocp}) of a metal changes when electrochemical reactions occur. Once equilibrium is reached, a steady value is measured, and then the potentiodynamic polarization curve is performed by providing ranges of potential values. The plots are then used to calculate the corrosion potential (E_{corr}) and the corrosion current density (i_{corr}). Additionally, different concentrations of green inhibitors and experimental temperatures can be used to examine their different effects on corrosion prevention performance [88].

Electrochemical impedance spectroscopy is one of the best and powerful analytical tools for following in situ electrochemical progression with insight into the physical phenomena acting at the metal-electrolyte interface, providing valuable information on the surface properties and electrode kinetics via impedance diagrams. In this technique, an AC voltage (in the case of potentiostatic EIS) or current (in the case of galvanostatic EIS) is applied to the system under study to receive a response in the form of AC current (voltage) or voltage (current) as a function of the frequency. This technique can be implemented in a three-electrode cell, similar to Potentiodynamic polarization. Electrochemical impedance spectroscopy is usually used to determine resistance and current flow values, both when green corrosion inhibitor is present in the solution and when it is not. The reported result is usually a Nyquist diagram, with the real part of the impedance (Z′) on the X-axis and the imaginary part (Z″) on the Y-axis [89].

The two main adsorption mechanisms are physisorption and chemisorption. It has been recommended that physisorbed molecules attach to the surface at the cathodes and basically retard metal dissolution by the cathodic reaction, whiles chemisorbed molecules shield anodic areas and reduce the inherent reactivity at the sites where they are attached. It is acknowledged that the values of the standard free energy of adsorption ΔG^{o}_{ads} in aqueous solution are −20 kJ.mol^{-1} or lower establishes the physisorption process. While, those around −40 kJ.mol^{-1} or more negative include charge sharing or transfer of electrons from inhibitory molecules to the metal surface, point toward coordinate or covalent bond. It is important to signal that both mechanisms can take place together on the same metal surface. Isotherm equations were used to validate the adsorption mechanism, and to ascertain the closest equation that relates the dosage of inhibitors to the adsorbed concentration at saturation. Empirical equations functioning as hyperbolic, exponential, logarithmic, and power are complicated to relate to the given adsorption mechanisms. There are many mathematical equations called adsorption models that estimate the adsorbate amount in the absorbent at constant temperature. Most of the time, green inhibitors obey the Langmuir isotherm model, but some also adhere to the Freundlich and Frumkin isotherms [7].

However, the study of the precise mechanism of the adsorption process is complex because most of the constituents moderate the corrosion reactions in many ways, which makes it difficult to assign the credit for corrosion mitigation to a particular constituent. Moreover, the nature of the adsorption of an inhibitor onto a metal surface is largely governed by characteristics such as chemical and electronic properties of the inhibitor, temperature, type of electrolyte, steric effects, and the nature and charge of the surface of the metals [90]. The negative surface charge will enhance the adsorption of the cation while the adsorption of the anion with the positive surface charge is preferred.

Simulation and computational modeling backed by wet experimental results would help to better understand the mechanism of inhibitor action, their adsorption patterns, and the inhibitor metal surface interface and aid the development of designer inhibitors with an understanding of the time required for the release of self-healing inhibitors. This is achieved by density functional theory (DFT) which is based on quantum chemical calculations that have emerged as potential tools for studying metal-inhibitor interactions between inhibitors and metallic surfaces [91]. Monte Carlo simulation is also well known as a traditional and powerful method if computational complexity and time are not limiting [92].

5. General extraction methods and challenges faced for sustainability

The most crucial step is the field of green inhibitors is the extraction of active substances. In general, extraction is a separation process, in which the active ingredients are isolated from the plant. Proper extraction processes are needed to extract the required active ingredients from the plants. The suitability of the extraction method depends on the polar or non-polar nature of the target compound, the size of the sample particles, and the presence of interfering materials. In addition, the selection of the solvent for extraction has to be rigorous and guarded as it should be based on the type of plant, part to be extracted, the availability of solvent, and the nature of the bioactive compounds [7]. Generally, extraction of polar compounds involves polar solvents, while nonpolar solvents are applied in extraction of nonpolar compounds. Several commonly used techniques [7, 91] can be applied to separate and extract the required extract from the plants among them:

- Supercritical fluid extraction
- Microwave-assisted extraction
- Ultrasound-assisted extraction
- Soxhlet extraction
- Enzyme assisted extraction
- Hydro-distillation extraction
- Steam-distillation extraction
- Ultra-high pressure extraction
- Accelerated solvent extraction, etc.

Each extraction method has its own characteristics. The type of extraction process can greatly affect the final natural products obtained. It should be chosen with caution according to the objective of the study. It has an effect on purity, price and yield and depends on the compound of interest and the required degree of purity. However, the choice of extraction and purification methods is another important fact. Some process is tedious, cumbersome, energy-consuming, time-consuming, and expensive. For example, plant extracts comprise only tiny portion of active constituents, therefore, a large amount of plants is mandatory to achieve satisfactory inhibition ability which results in a high cost. High temperature could lead to the deterioration of the sensitive active constituents and thus reduce the relative inhibition efficiency. In addition, the extraction process is too complex to be appropriate for large-scale applications in industries. Extraction requires relatively the use of high-level organic solvents that emit greenhouse gases that threatened humans, agriculture, and microorganisms. Moreover, excessive use of solvents leads to enormous waste of by-products. Unlike these dangerous techniques, further research is needed to introduce efficient and environmentally friendly processes such as "green processing," "green solvents," and "green products."

6. Conclusion

This chapter collects a variety of discussed and summarized studies advocating the use of plant extracts, oils, rare earths, and amino acids as corrosion inhibitors of metals and alloys. These inhibitors also present certain challenges, but they offer many advantages, and they remain the most ideal and promising alternative as they are generally synthesized from natural and non-toxic products, which increases their availability and effectiveness in terms of environmental and human safety.

Conflict of interest

The authors declare no conflict of interest.

Author details

Malak Rehioui
Laboratory of Organic Chemistry, Catalysis and Environment, Chemistry Department, Faculty of Sciences, Ibn Tofail University, Kenitra, Morocco

*Address all correspondence to: rehioui.malak0@gmail.com

References

[1] Zehra S, Mobin M, Aslam J. An overview of the corrosion chemistry. Environmentally Sustainable Corrosion Inhibitors. 2022:3-23

[2] Chigondo M, Chigondo F. Recent natural corrosion inhibitors for mild steel: An overview. Journal of Chemistry. 2016;**2016**:1-7

[3] Birat J-P. Chapter 5 Corrosion and oxidation of materials. Sustainable Materials Science—Environmental Metallurgy. 2020;**1**:229-246

[4] Koch G. Cost of corrosion. Trends in Oil and Gas Corrosion Research and Technologies. 2017:3-30

[5] Hansson CM. The impact of corrosion on society. Metallurgical and Materials Transactions A: Physical Metallurgy and Materials Science. 2011;**42**:2952-2962

[6] Zehra S, Mobin M, Aslam R. Chapter 2—Corrosion prevention and protection methods. Eco-Friendly Corrosion Inhibitors. 2022:13-26

[7] Kumari P, Lavanya M. Plant extracts as corrosion inhibitors for aluminium alloy in NaCl environment—Recent review. Journal of the Chilean Chemical Society. 2022:67

[8] Abbout S. Green inhibitors to reduce the corrosion damage. In: Singh A, editor. Corrosion. London: IntechOpen; 2020

[9] Abo El-Enin SA, Amin A. Review of corrosion inhibitors for industrial applications. International Journal of Engineering Research and Reviews. 2015;**3**:127-145

[10] Goni LKMO, Mazumder MAJ. Green Corrosion Inhibitors. IntechOpen: Corrosion inhibitors; 2019

[11] Shehata OS, Korshed LA, Attia A. Green Corrosion Inhibitors, Past, Present, and Future. IntechOpen: Corrosion Inhibitors; 2018

[12] Shahid M. Corrosion protection with eco-friendly inhibitors. Advances in Natural Sciences: Nanoscience and Nanotechnology. 2010;**2**:1-6

[13] Singh WP, Bockris JOM. Toxicity issues of organic corrosion inhibitors: Applications of QSAR model. Nace Corrosion. 1996

[14] Obot IB, Obi-Egbedi NO, Umoren SA, Ebenso EE. Synergistic and antagonistic effects of anions and ipomoea invulcrata as green corrosion inhibitor for aluminium dissolution in acidic medium. International Journal of Electrochemical Science. 2010;**5**:994-1107

[15] Amitha Rani BE, Basu BBJ. Green inhibitors for corrosion protection of metals and alloys: An overview. International Journal of Corrosion. 2012;**2012**:1-15

[16] Boxall ABA. Global climate change and environmental toxicology. In: Wexler P, editor. Encyclopedia of Toxicology. 3rd ed. Elsevier Science; 2014. pp. 736-740

[17] Umoren SA, Solomon MM, Obot IB, Suleiman RK. A critical review on the recent studies on plant biomaterials as corrosion inhibitors for industrial metals. Journal of Industrial and Engineering Chemistry. 2019;**76**:91-115

[18] Sabirneeza AAF, Geethanjali R, Subhashini S. Polymeric corrosion inhibitors for iron and its alloys: A review. Chemical Engineering Communications. 2015;**202**:232-244

[19] Kobzar YL, Fatyeyeva K. Ionic liquids as green and sustainable steel corrosion inhibitors: Recent developments. Chemical Engineering Journal. 2021;**425**:131480

[20] Sliem MH, Afifi M, Bahgat Radwan A, Fayyad EM, Shibl MF, Heakal FE-T, et al. AEO7 surfactant as an eco-friendly corrosion inhibitor for carbon steel in HCl solution. Scientific Reports. 2019;**9**:2319

[21] Hinton BRW, Arnott DR, Ryan NE. Inhibition of aluminum alloy corrosion by cerous cations. Metals Forum. 1984;7:211-217

[22] Hinton BRW. Corrosion inhibition with rare earth metal salts. Journal of Alloys and Compounds. 1992;**180**:15-25

[23] Somers AE, Hinton BRW, de Bruin-Dickason C, Deacon GB, Junk PC, Forsyth M. New, environmentally friendly, rare earth carboxylate corrosion inhibitors for mild steel. Corrosion Science. 2018;**139**:430-437

[24] Manh TD, Hien PV, Nguyen QB, Quyen TN, Hinton BRW, Nam ND. Corrosion inhibition of steel in naturally-aerated chloride solution by rare-earth 4-hydroxycinnamate compound. Journal of the Taiwan Institute of Chemical Engineers. 2019;**104**:177-189

[25] Peng Y, Hughes AE, Deacon GB, Junk PC, Hinton BRW, Forsyth M, et al. A study of rare-earth 3-(4-methylbenzoyl)-propanoate compounds as corrosion inhibitors for AS1020 mild steel in NaCl solutions. Corrosion Science. 2018;**145**:199-211

[26] Zhao D, Sun J, Zhang L, Tan Y, Li J. Corrosion behavior of rare earth cerium based conversion coating on aluminum alloy. Journal of Rare Earths. 2010;**28**:371-374

[27] Porcayo-Calderon J, Ramos-Hernandez JJ, Porcayo-Palafox E, de la Escalera LMM, Canto J, Gonzalez-Rodriguez JG, et al. Sustainable development of corrosion inhibitors from electronic scrap: Synthesis and electrochemical performance. Advances in Materials Science and Engineering. 2019;**2019**:1-14

[28] Bregman JI. Corrosion Inhibitors. New York: Macmillan; 1963

[29] Evans UR. The Corrosion and Oxidation of Metals. London: Edward Arnold; 1960. pp. 170-178

[30] Kaur J, Daksh N, Saxena A. Corrosion inhibition applications of natural and eco-friendly corrosion inhibitors on steel in the acidic environment: An overview. Arabian Journal for Science and Engineering. 2022;**47**:57-74

[31] Wei G, Deng S, Li X. Eupatorium Adenophora (Spreng.) leaves extract as a highly efficient eco-friendly inhibitor for steel corrosion in trichloroacetic acid solution. International Journal of Electrochemical Science. 2022;**17**:1-20

[32] Abeng FE, Idim VD. Green corrosion inhibitor for mild steel in 2 M HCl solution: Flavonoid extract of Erigeron floribundus. World Scientific News. 2018;**98**:89-99

[33] Fouda AS, Hegazi MM, El-Azaly A. Henna extract as green corrosion inhibitor for carbon steel in hydrochloric acid solution. International Journal of Electrochemical Science. 2019;**14**:4668-4682

[34] Rehioui M, Abbout S, Benzidia B, Hammouch H, Erramli H, Ait Daoud N, et al. Corrosion inhibiting effect of a green formulation based on Opuntia Dillenii seed oil for iron in acid rain solution. Heliyon. 2021;7:e06674

[35] Torres-Acosta AA. Opuntia-Ficus-Indica (Nopal) mucilage as a steel corrosion inhibitor in alkaline media. Journal of Applied Electrochemistry. 2007;**37**:835-841

[36] Belakhdar A, Ferkous H, Djellali S, Sahraoui R, Lahbib H, Ben Amor Y, et al. Computational and experimental studies on the efficiency of *Rosmarinus officinalis* polyphenols as green corrosion inhibitors for XC48 steel in acidic medium. Colloids and Surfaces A: Physicochemical and Engineering Aspects. 2020;**606**:125458

[37] Chellouli M, Chebabe D, Dermaj A, Erramli H, Bettach N, Hajjaji N, et al. Corrosion inhibition of iron in acidic solution by a green formulation derived from Nigella sativa L. Electrochimica Acta. 2016;**204**:50-59

[38] Elgyar OA, Ouf AM, El-Hossiany A, Fouda AE. The inhibition action of Viscum Album extract on the corrosion of carbon steel in hydrochloric acid solution. Biointerface Research in Applied Chemistry. 2021;**11**:14344-14358

[39] Batah A, Anejjar A, Bammou L, Belkhaouda M, Salghi R. Effect of apricot almond oil as green inhibitor for steel corrosion in hydrochloric media. Portugaliae Electrochimica Acta. 2020;**2020**(38):201-214

[40] Zouarhi M, Chellouli M, About S, Hammouch H, Dermaj A, Said Hassane SO, et al. Inhibiting effect of a green corrosion inhibitor containing Jatropha Curcas seeds oil for iron in an acidic medium. Portugaliae Electrochimica Acta. 2018;**36**:179-195

[41] Abbout S, Zouarhi M, Chebabe D, Damej M, Berisha A, Hajjaji N. Galactomannan as a new bio-sourced corrosion inhibitor for iron in acidic media Said. Heliyon. 2020;**6**:e03574

[42] Souza AV, da Rocha JC, Ponciano Gomes JAC, Palermo LCM, Mansur CRE. Development and application of a passion fruit seed oil microemulsion as corrosion inhibitor of P110 carbon steel in CO2-saturated brine. Colloids and Surfaces A. 2020;**599**:124934-124948

[43] Nasr K, Fedel M, Essalah K, Deflorian F, Souissi N. Experimental and theoretical study of *Matricaria recutita* chamomile extract as corrosion inhibitor for steel in neutral chloride media. Anti-Corrosion Methods and Materials. 2018;**65**:292-309

[44] Sahooa S, Nayaka S, Sahoo D, Mallik M. Corrosion inhibition behavior of dual phase steel in 3.5 wt % NaCl solution by Carica papaya peel extracts. Materials Today: Proceedings. 2019;**18**:2642-2648

[45] Singh A, Ahamad I, Quraish MA. *Piper longum* extract as green corrosion inhibitor for aluminium in NaOH solution. Arabian Journal of Chemistry. 2016;**9**:S1584-S1589

[46] Ahmed RK, Zhang S. Bee pollen extract as an eco-friendly corrosion inhibitor for pure copper in hydrochloric acid. Journal of Molecular Liquids. 2020;**316**:113849

[47] Abdelaziz S, Benamira M, Messaadia L, Boughoues Y, Lahmar H, Boudjerda A. Green corrosion inhibition of mild steel in HCl medium using leaves extract of *Arbutus unedo* L. plant: An experimental and computational approach. Collid Surface A. 2021;**619**:126496

[48] Xu C, Tan B, Zhang S, Li W. Corrosion inhibition of copper in sulfuric acid by *Leonurus japonicus* Houtt. Extract as a green corrosion inhibitor: Combination of experimental and theoretical research. Journal of the Taiwan Institute of Chemical Engineers. Chemical Engineers. 2022;**139**:104532

[49] Shahen S, Abdel-karim AM, Gaber GA. Eco-friendly Roselle (*Hibiscus sabdariffa*) leaf extract as naturally corrosion inhibitor for Cu-Zn alloy in 1M HNO3. Egyptian Journal of Chemistry. 2022;**65**:351-361

[50] Gonzalez-Rodriguez JG, Gutierrez-Granda DG, Larios-galvez AK, Lopez-sesenes R. Use of thymus vulgaris extract as green corrosion inhibitor for bronze in acid rain. Journal of Bio- and Tribo-Corrosion. 2022;**8**:77

[51] Deyab MA, Al-Qhatani MM. Green corrosion inhibitor: *Cymbopogon schoenanthus* extract in an acid cleaning solution for aluminum brass. Zeitschrift für Physikalische Chemie. 2021;**236**:215-226

[52] Ansari A, Ou-Ani O, Oucheikh L, Youssefi Y, Chebabe D, Oubair A, et al. Experimental, theoretical modeling and optimization of inhibitive action of *Ocimum basilicum* essential oil as green corrosion inhibitor for C38 steel in 0.5 M H2SO4 medium. Chemistry Africa. 2022;**5**:37-55

[53] Najem A, Sabiha M, Laourayed M, Belfhaili A, Benhiba F, Boudalia M, et al. New green anti-corrosion inhibitor of citrus peels for mild steel in 1 M HCl: Experimental and theoretical approaches. Chemistry Africa. 2022;**5**:969-986

[54] Zhang X, Jiang W-F, Wang H-L, Hao C. Adsorption and inhibitive properties of Apigenin derivatives as eco-friendly corrosion inhibitors for brass in nitric acid solution. Journal of Adhesion Science and Technology. 2019;**33**:736-760

[55] Sun X, Qiang Y, Hou B, Zhu H, Tian H. Cabbage extract as an eco-friendly corrosion inhibitor for X70 steel in hydrochloric acid medium. Journal of Molecular Liquids. 2022;**365**:119733

[56] Wu Y, Zhang Y, Jiang Y, Qian Y, Guo X, Wang L, et al. Orange peel extracts as biodegradable corrosion inhibitor for magnesium alloy in NaCl solution: Experimental and theoretical studies. Journal of the Taiwan Institute of Chemical Engineers. 2020;**115**:35-46

[57] Onukwuli OD, Anadebe VC, Okafor C. Optimum prediction for inhibition efficiency of *Sapium ellipticum* leaf extract as corrosion inhibitor of aluminum alloy (AA3003) in hydrochloric acid solution using electrochemical impedance spectroscopy and response surface methodology. Bulletin of the Chemical Society of Ethiopia. 2020;**34**:175-191

[58] Boudalia M, Fernández-Domene RM, Tabyaoui MH, Bellaouchou A, Genbour A, García-AntónJ. Green approach to corrosion inhibition of stainless steel in phosphoric acid of Artemesia herba albamedium using plant extract. Journal of Materials Research and Technology. 2019;**8**:5763-5773

[59] Chraka A, Raissouni I, Seddik NB, Khayar S, Ibn Mansour A, Tazi S, et al. Identification of potential green inhibitors extracted from *Thymbra capitata* (L.) Cav. For the corrosion of Brass in 3% NaCl solution: Experimental, SEM–EDX analysis, DFT computation and Monte Carlo simulation studies. Journal of Bio- and Tribo-Corroions. 2020;**6**:80

[60] Hamadi L, Mansouri S, Oulmi K, Kareche A. The use of amino acids as corrosion inhibitors for metals: A review. Egyptian Journal of Petroleum. 2018;**2018**(27):1157-1165

[61] Tonouchi N, Ito H. Present global situation of amino acids in industry. Advances in Biochemical Engineering/Biotechnology. 2017;**159**:3-14

[62] El-Sayed NH. Corrosion inhibition of carbon steel in chloride solutions by some amino acids. European Journal of Chemistry. 2016;7:14-18

[63] Amin MA, Khaled KF. Copper corrosion inhibition in O2-saturated H2SO4 solutions. Corrosion Science. 2010;**52**:1194-1204

[64] Srivastava V, Haque J, Verma C, Singh P, Lgaz H, Salghi R, et al. Amino acid based imidazolium zwitterions as novel and green corrosion inhibitors for mild steel: Experimental, DFT and MD studies. Journal of Molecular Liquids. 2017;**244**:340-352

[65] Zeino A, Abdulazeez I, Khaled M, Jawich MW, Obot IB. Mechanistic study of polyaspartic acid (PASP) as eco-friendly corrosion inhibitor on mild steel in 3% NaCl aerated solution. Journal of Molecular Liquids. 2018;**250**:50-62

[66] Amin MA, Ibrahim M. Corrosion and corrosion control of mild steel in concentrated H2SO4 solutions by a newly synthesized glycine derivative. Corrosion Science. 2011;**53**:873-885

[67] Zhang D-Q, Cai Q-R, Gao L-X, Lee KY. Effect of serine, threonine and glutamic acid on the corrosion of copper in aerated hydrochloric acid solution. Corrosion Science. 2008;**50**:3615-3621

[68] Nady H. Tricine [N-(Tri (hydroxymethyl)methyl)glycine]—A novel green inhibitor for the corrosion inhibition of zinc in neutral aerated sodium chloride solution. Egyptian Journal of Petroleum. 2017;**26**:905-913

[69] Yeganeh M, Khosravi-Bigdeli I, Eskandari M, Alavi Zaree SR. Corrosion inhibition of l-methionine amino acid as a green corrosion inhibitor for stainless steel in the H2SO4 solution. Journal of Materials Engineering and Performance. 2020;**29**:3983-3994

[70] Jano A, Lame A, Kokalari E. Lysine as corrosion inhibitor for low alloy carbon steel in acidic media. Analele Universitatii "Ovidius" Constanta—Seria Chimie. 2014;**25**:11-14

[71] Kowsari E, Arman SY, Shahini MH, Zandi H, Ehsani A, Naderi R, et al. In situ synthesis, electrochemical and quantum chemical analysis of an amino acid-derived ionic liquid inhibitor for corrosion protection of mild steel in 1M HCl solution. Corrosion Science. 2016;**16**:73-85

[72] Gowri S, Sathiyabama J, Rajendran S, Robert Kennedy Z, Agila DS. Corrosion inhibition of carbon steel in sea water by glutamic acid-Zn2+ system. Chemical Science Transactions. 2013;**2**:275-281

[73] Gowri S, Sathiyabama J, Rajendran S. Corrosion inhibition of carbon steel in sea water by L-arginine-Zn2+ system. International Journal of Chemical Engineering. 2014;**2014**:1-9

[74] Wang D, Gao L, Zhang D, Yang D, Wang H, Lin T. Experimental and theoretical investigation on corrosion inhibition of AA5052 aluminium alloy by l-cysteine in alkaline solution. Materials Chemistry and Physics. 2015;**169**:142-151

[75] Fu J-J, Li S-N, Cao L-H, Wang Y, Yan L-H, Lu L-d. L-tryptophan as green corrosion inhibitor for low carbon steel in hydrochloric acid solution. Journal of Materials Science. 2010;**45**:979-986

[76] Prathipa V, Raja AS, Prabha SS. L-aspartic acid: An efficient water soluble inhibitor for corrosion of carbon steel in aqueous media. International Journal of Chemical, Material and Environmental Research. 2016;**3**:35-41

[77] Unnimaya, Shetty P, Kumari P, Kagatlkar S. Glutathione as green corrosion inhibitor for 6061Al-SiC(p)

composite in HCl medium: Electrochemical and theoretical investigation. Journal of Solid State Electrochemistry. 2023;**27**:255-270

[78] Sahaya Raja A, Rajendran S, Sathiyabama J, Prathipa V, Karthika IN, Krishnaveni A. Use of L-alanine as nature—Friendly corrosion inhibitor for carbon steel In aqueous medium. International Journal of Nano Corrosion Science and Engineering. 2015;**2**:26-40

[79] Zhang QH, Hou BS, Li YY, Lei Y, Wang X, Liu HF, et al. Two amino acid derivatives as high efficient green inhibitors for the corrosion of carbon steel in CO2-saturated formation water. Corrosion Science. 2021;**189**:109596

[80] Loto RT. Corrosion inhibition effect of non-toxic α-amino acid compound on high carbon steel in low molar concentration of hydrochloric acid. Journal of Materials Research and Technology. 2019;**8**:484-493

[81] Rahiman A, Subhashini S. Corrosion inhibition, adsorption and thermodynamic properties of poly(vinyl alcohol-cysteine) in molar HCl. Arabian Journal of Chemistry. 2017;**10**:S3358-S3366

[82] Raja A, Rajendran S, Satyabama P. Inhibition of corrosion of carbon steel in well water by DL-phenylalanine-Zn2+ system. Journal of Chemistry. 2013;**2013**:1-8

[83] Zunita M, Wahyuningrum D, Wenten I, Boopathy R. Carbon steel corrosion inhibition activity of tofu associated proteins. Bioresource Technology Reports. 2022;**17**:100973

[84] Wang J, Liu C, Qian B. A novel L-histidine based ionic liquid (LHIL) as an efficient corrosion inhibitor for mild steel. Royal Society of Chemistry Advances. 2022;**12**:2947-2958

[85] Singh P, Srivastava V, Quraishi MA. Novel quinoline derivatives as green corrosion inhibitors for mild steel in acidic medium: Electrochemical, SEM, AFM, and XPS studies. Journal of Molecular Liquids. 2016;**216**:164-173

[86] Issaadi S, Douadi T, Zouaoui A, et al. Novel thiophene symmetrical Schiff base compounds as corrosion inhibitor for mild steel in acidic media. Corrosion Science. 2011;**53**:1484-1488

[87] Verma C, Quraishi MA. Thermodynamic, electrochemical and surface studies of dendrimers as effective corrosion inhibitors for mild steel in 1 M HCl. Analytical and Bioanalytical Electrochemistry. 2016;**8**:104-123

[88] Popoola LT. Organic green corrosion inhibitors (OGCIs): a critical review. Corrosion Reviews. 2019;**37**:71-102

[89] Al-Amiery AA, Kadhum AAH, Kadihum A, Mohamad AB, How CK, Junaedi S. Inhibition of mild steel corrosion in sulfuric acid solution by new Schiff base. Materials (Basel). 2014;7:787-804

[90] Maayta AK, Al-Rawashdeh NAF. Inhibition of acidic corrosion of pure aluminum by some organic compounds. Corrosion Science. 2004;**46**:1129-1140

[91] Verma DK. Density Functional Theory (DFT) as a Powerful Tool for Designing Corrosion Inhibitors in Aqueous Phase. London, UK: Intechopen; 2018

[92] Verma CB, Lgaz H, Verma DK, Ebenso EE, Bahadur I, Quraishi MA. Molecular dynamics and Monte Carlo simulations as powerful tools for study of interfacial adsorption behavior of corrosion inhibitors in aqueous phase: A review. Journal of Molecular Liquids. 2018;**260**:99-120

Chapter 4

Organic Corrosion Inhibitors

Bharat Chandra Sahu

Abstract

Organic corrosion inhibitors are preferred due to its environmental friendly and effectiveness at a wide range of temperatures. The efficiency of an organic inhibitor depends on the size of the organic molecule, aromaticity, type, and number of bonding atoms or groups in the molecule (either π or σ), nature and surface charge, the distribution of charge in the molecule, and type of aggressive media. The presence of polar functional groups with S, O, or N atoms in the molecule, heterocyclic compounds and pi electrons present in the molecule also increases the efficiency of these organic corrosion inhibitors. The use of computational chemistry such as density functional theory (DFT), molecular dynamic simulation (MD), Monte Carlo (MC) simulations, and quantitative structure-activity relationship (QSAR) modeling has been applied for study of corrosion inhibition properties of organic compounds. This chapter will explain about theoretical and computational study of organic compounds as corrosion inhibitors.

Keywords: organic corrosion inhibitors, physisorption, cathodic inhibitors, anodic inhibitors, mixed inhibitors

1. Introduction

Corrosion is a natural process and also a universal phenomenon of metals. Corrosion is the deterioration of the metal in the presence of the surrounding environment by chemical or electrochemical means [1–10]. The corrosion of metals results into a huge loss of economy in the world [11–14]. There are various methods to reduce or prevent corrosion. The use of corrosion inhibitors is one of the ways to protect metal surfaces against corrosion [15–18]. In the early days, inorganic compounds such as nitrates, chromates, borates, silicates, chromates, phosphate, and molybdates were used as inhibitors for the prevention of corrosion [19–31]. Due to the toxic and environmental effects of that inorganic inhibitor, scientists start using environment friendly organic inhibitor for prevention of corrosion [32–35]. The organic compounds that contain mainly nitrogen, oxygen, sulfur phosphorus, and multiple bonds or aromatic rings in their structures act as corrosion inhibitors [36, 37]. Various heterocyclic aromatic compounds are studied as organic inhibitors [38–48]. It has been studied by various groups that carboxylate and amine of aliphatic compounds can be used as organic inhibitors [49–52].

In this article, there will be a detailed study of organic compounds as inhibitors and their computational and theoretical studies will be explained. There are various reviews papers on molecular modeling of compounds as corrosion inhibitors by

 IntechOpen

various groups. The DFT and MD s simulation studies about adsorption of the organic corrosion inhibitors and the metal surface in various corrosive environments.

2. Organic corrosion inhibitors

Organic corrosion inhibitors are preferred due to its environmental friendly and effectiveness at wide range of temperatures [53–59]. The efficiency of an organic inhibitor depends on the size of the organic molecule, aromaticity, type, and number of bonding atoms or groups in the molecule (either π or σ), nature and surface charge, the distribution of charge in the molecule and type of aggressive media. The presence of polar functional groups with S, O, or N atoms in the molecule, heterocyclic compounds and pi electrons present in the molecule also increases the efficiency of these organic corrosion inhibitors. The adsorption of the molecule on the metal surface depends on the polar function of the molecule. The organic compound that contains oxygen, nitrogen and/or sulfur blocked the active corrosion sites by adsorbing on the metallic surface.

The organic compounds act as cathodic, anodic, and mixed inhibitors. The organic molecules exhibiting a strong affinity for metal surfaces show good inhibition efficiency. The organic compounds adsorb on the surface of the metal to form a protective film which displace water from the metal surface and protect it against corrosion. The organic inhibitors adsorb onto the metallic surface depend on the chemical structure of the inhibitors, type of environment, and surface charge of the metal. The inhibitor adsorption on metal surfaces occurs through donor-acceptor interactions (physical, chemical, or mixed type) [60–63]. The organic inhibitors are adsorbed on metal surface either by physisorption or chemisorptions mechanism. The physisorption occurs through an electrostatic interaction between the metal surface and inhibitor's charged molecule [64–66]. The chemisorptions mechanism occurs through charge transfer share between the adsorbed inhibitor molecules and the metal surface.

The mechanism of organic corrosion inhibition takes place in two steps. The first step involves the transfer of the corrosion inhibitors over the metallic surface. In the second step interactions between metal surface and adsorbed inhibitor molecules occur [67–69]. In physisorption, the adsorbed inhibitor molecules are not in direct contact with the metallic surface. A layer of solvents which are already adsorbing on the surface molecules separated the inhibitor molecules [70, 71].

The mechanism of organic corrosion inhibition in aqueous medium can be occurs as follows [72].

$$M + nH_2O \leftrightarrow M(H_2O)_{nads} \tag{1}$$

$$M(H_2O)_{n\ ads} + 2X^- \leftrightarrow M\left[(H_2O)_n(X)_2^-\right]_{ads} \tag{2}$$

$$M\left[(H_2O)_n(X)_2^-\right]_{ads} \leftrightarrow M\left[(H_2O)_n(X)_2\right]_{ads} + 2e^- \tag{3}$$

$$M\left[(H_2O)_n(X)_2\right]_{ads} \leftrightarrow M^{2+} + OH^- + 2X^- + H^+ \tag{4}$$

In the presence of organic corrosion inhibitor (INH), the following mechanism operates [73, 74].

$$M + nH_2O \leftrightarrow M(H_2O)_{nads} \tag{5}$$

$$M(H_2O)_{n\ ads} + 2X^- \leftrightarrow M\left[(H_2O)_n(X)_2^-\right]_{ads} \tag{6}$$

$$M\left[(H_2O)_n(X)_2^-\right]_{ads} + INH - H^+ \leftrightarrow M\left[(H_2O)_n(X)_2INH - H\right]^-_{ads} \tag{7}$$

The effect of organic compounds as corrosion inhibitors and adsorption on metal surfaces depends on various factors. Temperature is one of the factors. The increase in temperature results in a decrease in the inhibition of corrosion due to adsorption on metallic surface decreases [75, 76]. Then inhibitor concentration is another parameter for the effectiveness of the organic inhibitor molecules. With increasing inhibitor concentration, inhibition efficiency of compounds increases. However, after certain concentration, there is no further enhancement of inhibition. The effectiveness of the organic inhibitor also depends on the electronic structure and molecular size of the molecule [77, 78]. The organic compound with high molecular size generally shows a better corrosion inhibitor than compound with lower molecular size. But, too big molecular size compound show a decrease in adsorption on metallic surfaces due to a decrease in solubility. The molecular electronic structure is one of the aspects of corrosion inhibitors. The planar geometry organic compound acts as better corrosion inhibitor than with vertical geometry of the compounds. The aromatic organic compounds having conjugation in the form behave as effective corrosion inhibitors The aromatic compound containing electron-donating substituent such as $-OCH_3$, $-CH_3$, $-OH$, $-NH_2$, $-NHR$, and $-NR_2$ increases the organic corrosion inhibitor molecules, whereas, electron-withdrawing substituent such as $-CN$, $-COOH$, $-COOC_2H_5$, and $-NO_2$ decreases inhibitor molecules [79]. The effect of aromatic substituted substituent on corrosion inhibition can be determined by Hammett substituent constant (σ) values. The corrosion inhibition efficiency of substituted aliphatic linear and cyclic compounds can be determined by Taft constant (σ^*) values [80]. The Hammett equation is shown below

$$log = \frac{K_R}{K_H} = \rho\sigma \tag{8}$$

whereas K_H and K_R are equilibrium constants for non-substituted and substituted compounds. σ is the Hammett constant, ρ is its magnitude depending on the nature of metal-inhibitor interactions.

3. Computational method used for corrosion inhibitors

There are various experimental methods for the evaluation of the inhibition performance of corrosion inhibitors. The synthesis of organic compounds for corrosion inhibitors using experimental methods is expensive and time-consuming. The synthesis of organic inhibitors requires multiple steps and also so many side products are generated in the synthesis of organic molecules. Recently, computational methods have been used extensively for the study of corrosion inhibition [81–95]. The corrosion inhibition can be known by knowing the mechanism of interaction between the inhibitor molecules and the metal surface. The interaction of the metal and inhibitor at the atomic and molecular levels cannot be understood by experimental study. The deeper insights into the mechanism of corrosion inhibitors on metal surfaces can be understood by molecular modeling method [96]. Molecular modeling techniques provide mechanistic processes at the atomic and molecular levels [97–118].

The behaviors of atomic and molecular interactions at microscopic and macroscopic can be analyzed by molecular modeling techniques [119, 120]. The interactions between molecules at the microscopic level can be known by molecular modeling simulations. Molecular modeling techniques are divided into three categories: ab initio electronic structure methods, semi-empirical methods, and atomistic simulation.

The ab initio method is also called first principles electronic structure method and is based on the law of quantum mechanics. The different ab initio methods are the Hartree–Fock Self-Consistent Field (HF-SCF) method, Møller–Plesset perturbation theory (MPn), coupled cluster (CC), and density functional theory (DFT). The density functional theory (DFT) has been used widely for understanding metal-inhibitor interactions at the molecular level.

3.1 Density functional theory

DFT has been used for understanding the molecular structural behavior of corrosion inhibitors [121–125]. In the DFT method, one function is determined in terms of another function, which is essentially the meaning of the word "functional". The energy of a system in the DFT method is obtained from its electron density. Commonly used DFT methods are the B3LYP, BLYP, B3P86, B3PW91, and PW91. Basis sets are sets of linear combinations of mathematical functions that are used to describe the shapes of atomic orbitals. The use of basis sets is necessary to be able to carry out ab initio calculations. Basis sets describe atomic orbitals by assigning a group of basis functions to each atom within a molecule. There exist broad lists basis sets that are used to perform ab initio calculations. The DFT/ B3LYP method using the 6-31G and 6-311G basis sets is the most widely used in corrosion inhibition studies.

The mechanism of corrosion inhibition can be understood by using DFT simulations. DFT simulation helps in finding the interactions of inhibitor with metal surface. For understanding the interactions of inhibitor with metal surface, the computational methods provide information about highest occupied (HO) and lowest unoccupied (LU) molecular orbitals (MO), frontier orbital energies, energy band gap, hardness, electronegativity, Mulliken and Fukui population analyses, electron-donating power, electron-accepting power, chemical potential, hardness, softness, dipole moment, and number of electrons transferred.

3.1.1 HOMO and LUMO

The adsorption ability and corrosion inhibition effectiveness of a compound can be linked to energy of frontier molecular orbital (FMOs; E_{HOMO} and E_{LUMO}), hardness (Z), electronegativity (w), dipole moment (m), softness (s), and fraction of electron transfer (DN). The chemical relativities of molecules in a corrosion inhibitor molecule are related to HOMO and LUMO electron densities. The highest occupied molecular orbital (HOMO) donates electrons to the free d orbital of a metal. The lowest unoccupied molecular orbital (LUMO) accepts electrons from the metal. Lower values indicate higher tendency of accepting electrons. The positive values are connected with chemisorptions, whereas negative values are with physisorption. The higher electron-donating capability is associated with a higher HOMO in the heteroatom molecules heteroatom.

HOMO and LUMO were associated with electron-donating and electron-accepting abilities of molecules, respectively, according to FMOs theory. Inhibitor molecules having high energy of HOMO will be effective to transfer the electrons to a metallic surface and low LUMO energy value shows that the molecule is a good electron acceptor. Koopmans theorem [126] is a bridge between DFT and MO theory and it can be used in the prediction of ionization potential (IP) and electron affinity (EA) values of molecules. According to this theorem, IP and EA can be expressed via the following equations:

$$IP = -E_{HOMO} \tag{9}$$

$$EA\ E = -E_{LUMO} \tag{10}$$

Further, the energy difference between LUMO and HOMO called an energy gap (ΔE) is also an essential parameter toward the description of reactivity of a molecule.

$$\Delta E = E_{LUMO} - E_{HOMO} \tag{11}$$

ΔE having a large value show low reactivity of molecules with metal surface while a molecule with a low value of ΔE strongly adsorbed on a metal surface.

3.1.2 Electronegativity (η), chemical potential (μ), hardness (η), and softness (σ) indices

The η, μ, η, and σ parameters are related to the total electronic energy (E) with respect to the number of electrons (N) at a constant external potential. The η is defined as the negative value of μ. Within the framework of finite differences approaches, these parameters can be expressed in the form of ground-state IP and ground-state EA values of a chemical compound. The theoretical formulas can be expressed as [127]:

$$\mu = -\frac{IP + EA}{2} \tag{12}$$

$$\eta = -\frac{IP - EA}{2} \tag{13}$$

$$\sigma = -\frac{1}{\eta} \tag{14}$$

The electron-donating power and electron-accepting power of the molecule to accept and donate electrons are related to IP and EA. The inhibition abilities of the molecule based on their ability to accept and receive electrons provide information.

3.1.3 The fraction of electrons transferred (ΔN)

The tendency of an inhibitor molecule to transfer its electron to a metal surface, the hardness, and electronegativity predict ΔN. The interaction between the metal surface and inhibitor molecule on the basis of the fraction of electrons transferred. An inhibitor can transfer its electron if $\Delta N > 0$ and vice versa if ΔN [128–130].

3.1.4 Fukui indices (FIs)

The local reactivity and selectivity of molecule can be understood by Fukui functions. The nucleophilic and electrophilic regions of attack of inhibitor molecules are provided by Fukui function. The Fukui Indices (FIs) pinpoint the reactive sites in which the electrophilic or nucleophilic attacks are large or small. HSAB theory gave the prediction and interpretation of many CQ parameters and Fukui functions is also an early attempt in this direction Fukui Indices f(r) is the first derivative of ρ(r) with respect to the number of electrons (N) at a constant external potential v(r). FIs were identified with respect to hard or soft reagents by involving the HSAB principle. A simple approximation can be used with the aid of finite difference approximation and Mulliken's population analysis in which FIs were determined [131].

3.2 Atomistic simulations

Atomistic simulations also called force field methods or molecular mechanics are based on the principles of classical physics. The atomistic simulations are the investigation and simulation of physical phenomena on a molecular level. The two well-known atomistic simulation methods are the molecular dynamics (MD) [132–135] and Monte Carlo (MC) [136, 137] simulation techniques.

3.2.1 Molecular dynamics (MD) simulations

MD simulations provide the actual trajectory of a system by simulating the time evolution of the system. The concept of MD simulation is based on solving Newton's equations of motion for the atoms in the simulation system using numerical integration [138, 139].

3.2.1.1 Total energy minimization

The first step in atomistic simulations is to find a stable structure by the total energy minimization process (geometry optimization). The lowest energy configuration of the atoms in a system can reach by adjusting their coordinates. The net forces acting on the atoms are zero.

3.2.1.2 Ensemble

An ensemble is defined by its thermodynamic states such as constant pressure, temperature, and volume mimicking experimental conditions during a molecular mechanics simulation. The main ensembles used for the molecular dynamics simulation are the microcanonical ensemble (NVE), canonical ensemble, isobaric-isothermal (NPT), and grand canonical ensembles (μVT) [140].

3.2.1.3 Force fields

The set of parameters acting on the nuclei of atoms by which the potential energy (U) of a system is calculated of a molecular system is called Force fields. To find a correct force field is a challenging task in every simulation for a given system. Condensed-phase Optimized Molecular Potentials for Atomistic Simulation Studies (COMPASS) is the general force fields used for corrosion inhibition studies. In addition to ab initio calculations, COMPASS was parameterized considering various experimental data including organic compounds made with H, C, N, O, S, and P atoms, halogens, and metals. Universal force field (UFF) [141, 142], is an all-atom potential containing parameters for each atom, and the consistent-valence force field (CVFF) t is a generalized valence force field that allow treating of organic molecules as well as metals [143–145].

3.2.1.4 Periodic boundary condition

Atomistic simulations are usually carried out in a stimulation box. It can be obtained by treating a small system of about 100–10,000 particles using the periodic boundary condition approach. The corrosion inhibition systems are infinite systems where thousands of atoms and molecules are involved. The periodic boundary condition eliminates surface effects caused by the finite size of a system, and makes it more of an infinite one. In periodic boundary conditions, all atoms in the simulation box are replicated throughout space to form an infinite lattice.

3.2.2 Monte Carlo (MC) simulations

The MC method can simulate a system under thermodynamic equilibrium [146]. MC has the advantage of probabilistic investigation of the equilibrium behavior of systems as a function of temperature (Metropolis Monte Carlo) and advances the state of reactive systems through time (Kinetic Monte Carlo) [147]. The Metropolis MC, kinetic Monte Carlo (kMC), and quantum Monte Carlo are used in many physical and chemical applications based on the concept of Monte Carlo. Metropolis MC is simple and mostly used to study the interaction between inhibitor molecules and surface of metals in corrosion inhibition.

3.2.2.1 Interaction and binding energies

The interaction energy is defined as the required energy for one mole of an inhibitor molecule to be adsorbed on a metal surface [148]. For a simulated system in a vacuum, it can be determined using the following equation [149, 150]:

$$E_{inter} = E_{Total} - (E_{Surface} + E_{inh}) \tag{15}$$

In the presence of a solvent:

$$E_{inter} = E_{Total} - (E_{Surface+solvent} + E_{inh}) \tag{16}$$

where E_{Total}, $E_{Surface}$, $E_{Surface\ +\ solution}$, and E_{inh} denote the total energy of the simulated system, surface without solution, surface with solution, and inhibitor molecule alone, respectively.

The binding energy is defined as the negative values of the interaction energy. A large binding energy implies that the inhibitor molecule can be strongly adsorbed over a metal surface [151, 152].

$$E_{binding} = -E_{inter} \tag{17}$$

A representative case where interaction and binding energies are used to discuss the adsorption behavior of corrosion inhibitors has been performed by Lgaz et al. [153].

4. Aliphatic compounds as an organic corrosion inhibitor

Jia-jun Fu et al. carried out experimental, MD simulations and DFT calculations to investigate the corrosion inhibition properties of four amino acid compounds namely L-cysteine, L-histidine, L-tryptophan, and L-serine amino acid for mild steel in hydrochloric acid solution [154]. The optimization of the amino acid compounds has been performed using B3LYP Density Functional Theory (DFT) with 6-311G (d,p) basis set with the Gaussian 03 W program. This basis set will provide E_{HOMO}, E_{LUMO}, ΔE, and μ. The Fukui function calculations were performed with DMol3 version 4.3 available in Material Studio software (Accelrys, San Diego, CA), using a PBE (Perdew, Burke, and Enzerhof) functional and a double-numeric quality basis set with polarization functions (DND). Dmol3 uses a Mulliken population analysis.

The optimized structures of the four amino acid compounds were shown in **Figure 1**. Additionally, the orbital density distributions of E_{HOMO} and E_{LUMO} for amino acid compounds were shown in **Figure 2**. From this DFT calculation of these four amino acids, the E_{HOMO}, ΔE and ΔN provide that the order of the inhibition efficiency of these inhibitors follows the sequence: L-Try>L-His>L-Cys > L-Ser.

Mohammed et al. [155] carried out a study of the inhibition of iron corrosion in HCl solutions by three amino acids, namely alanine (Ala), cysteine (Cys), and S-methyl cysteine (S-MCys). The effectiveness of amino acids, alanine (Ala), cysteine (Cys), and S-methyl cysteine (S-MCys) as safe corrosion inhibitors for iron in aerated stagnant 1.0 M HCl solutions have been evaluated by Tafel polarization and

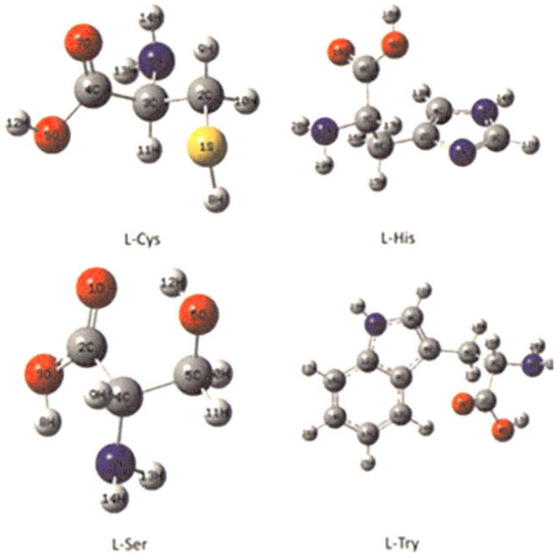

Figure 1.
Optimized structures of four amino acid compounds.

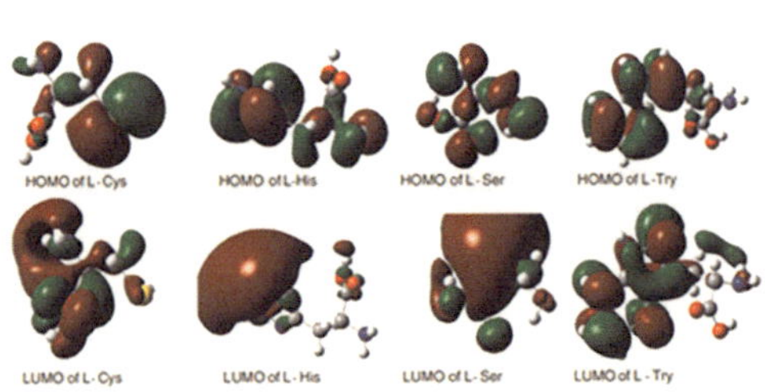

Figure 2.
The frontier molecular orbital distribution of four amino acids compounds.

impedance measurements. They have also performed molecular dynamics (MD) and density functional theory (DFT) for their inhibition efficiency. The electronic density distribution of HOMO (the highest occupied molecular orbital) and LUMO (the lowest unoccupied molecular orbital) for the studied amino acids were shown in **Figure 3.** From these studies, Cys has the lowest E_{HOMO} and the highest E_{LUMO} then S-MCys and Ala. The experimental data and computational study has been in agreement that Cys has a higher inhibition efficiency than alanine (Ala) and S-methyl cysteine (S-MCys).

Obot et al. studied the mechanism of 2-mercaptobenzimidazole adsorption on Fe (110), Cu (111), and Al (111) surfaces by applying DFT and molecular dynamics simulations [156]. The DFT study was performed using B3LYP, 6-31G (d,p) basis set in the gas phase. The inhibition was performed by MD simulations, optimization using the COMPASS force field under the NVT ensemble and PBC at a simulation time of 50 ps. The mechanism of corrosion inhibition of 2-mercaptobenzimidazole (2-MBI) on Fe, Cu, and Al surfaces Density functional theory (DFT) and molecular dynamics (MD) simulations.

Canul and Rosado studied the Inhibition Effect of Sodium Glutarate toward Carbon Steel Corrosion in Neutral Aqueous Solutions [157]. They investigated the Inhibition Effect of Sodium Glutarate in a near-neutral 0.02 M NaCl solution at ambient temperature for a range of concentrations (1–100 mM) by potentiodynamic polarization and electrochemical impedance spectroscopy (EIS) measurements. Sodium glutarate showed a poor inhibitive action for corrosion of carbon steel in a 0.02 M NaCl solution when used in concentrations of 1 mM and 5 mM. However, open circuit potential and polarization curve measurements give evidence that full chemical passivation is accomplished for concentrations of 32 mM or higher; a significant improvement in protective is achieved. Investigation of the effect of temperature showed that increasing the temperature from 22 to 55°C decreases the inhibition efficiency from 96 to 89%, indicating good stability of the protective film in this temperature range.

Ambat and co-workers reported electrochemical and molecular modeling studies of CO_2 corrosion inhibition characteristics of alkanolamine molecules (ethanolamine, diethanolamine, and triethanolamine) for the protection of 1Cr steel [158]. The DFT

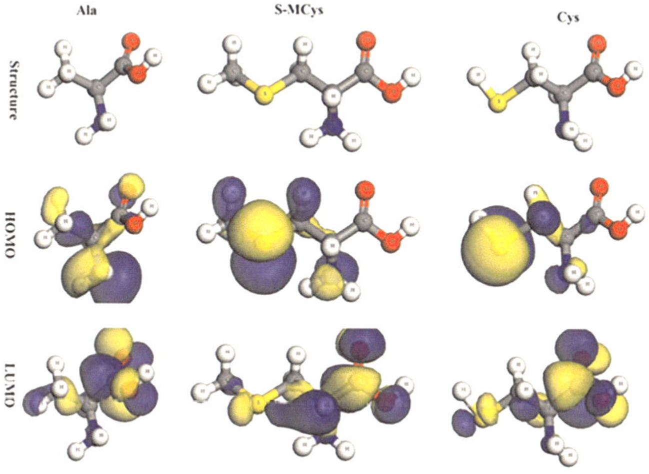

Figure 3.
Structure of Ala, S-Myc, and Cys.

calculations were performed with the Quantum Espresso package on a Fe (110) surface and a $FeCO_3$ (104) surface. The experimental results and molecular modeling calculations using DFT found that the inhibitor efficiency for the ethanolamine on Fe(110), $FeCO_3$(104), and Fe_3C(001) was the highest.

Obot and co-workers performed density functional theory and molecular dynamics simulation for study the corrosion inhibition for three amine derivatives, namely, N1-(2-aminoethyl)ethane-1,2-diamine (DETA), N1-(2-(2-aminoethylamino)ethyl) ethane-1,2-diamine (TETA) and N1-(2-(2-(2-(2-aminoethylamino)ethylamino) ethylamino)ethyl)ethane-1,2-diamine (PEHA) on the steel surface [159]. They have calculated E_{HOMO}, E_{LUMO}, energy gap (ΔE), electron affinity (A), electronegativity (χ), global hardness (η), softness (σ), and the fraction of electron transferred (ΔN) from the inhibitor molecule to the metal surface. The quantum chemical parameters E_{HOMO}, E_{LUMO}, ΔE, hardness (η), softness (σ), and fraction of electron transferred (ΔN) are in good agreement with the experimental findings. The nucleophilic and electrophilic characteristic of inhibitor molecules were analyzed from Fukui indices. MD simulation performed by COMPASS force field. The interaction energies at 298 K temperature are −355.30, −455.93, and − 784.74 kJ/mol for DETA, TETA, and PEHA, respectively, on the Fe (1 1 0) surface. PEHA has maximum interaction energy value and possesses better inhibition effectiveness than DETA and TETA. The corrosion inhibition property increases from DETA to PEHA. The binding energy values of inhibitor molecules are in the order PEHA > TETA > DETA which also supports the experimentally obtained results.

Bingul and co-workers studied the inhibition effects of methionine and tyrosine on the corrosion of Iron in HCl Solution by electrochemical and quantum-chemical method [160]. The quantum chemistry calculation for inhibition of corrosion has been performed with the 6-311G(d,p) basis set for methionine and tyrosine. The quantum chemical calculations were performed for methionine and tyrosine, as well as, for their protonated structures. The inhibition effect of methionine ion on Fe in HCl solution is better than that of tyrosine and is in agreement with experimental and theoretical data.

Kumari and Kumar have reported experimental and theoretical investigations of 3,3′-diamino dipropyl amine: highly efficient corrosion inhibitor for carbon steel in 2 N HCl at normal and elevated temperatures [161]. The theoretical study were performed by B3LYP using 6-31G(d) basis set and with the help of Gaussian 09, 6 Wallingford CT, the USA software. The adsorption of DADPA on carbon steel has been found from the data on global hardness, electronegativity, softness, chemical potential, ionization potential, maximum charge transfer, electron transfer, electrophilicity, nucleophilicity, electron releasing and accepting tendency, work function, back donation energy, proton affinity, metal inhibitor interaction, and binding energies. The strong adsorption of the inhibitor molecule was further supported by the suitable linear molecular configuration of DADPA, the high value of electronegativity, global softness parameter, and high negative value of interaction parameter.

Wang et al. studied experimental and theoretical investigation on corrosion inhibition of AA5052 aluminum alloy by L-cysteine in alkaline solution [162]. AA5052 aluminum alloy contains Mg, Mn, and Zn. The inhibition efficiencies of L-cysteine in 4 M NaOH solution were studied by polarization curves, electrochemical impedance spectroscopy, and quantum chemical calculation. The DFT calculations were performed using the Dmol3 package with numerical atomic orbital basis sets on Al (111) surface, double-numerical plus polarization (DNP) basis sets were used in the expansion of the Kohn-Sham orbitals and the orbital confining cut-off was set as 4.8 Å.

The adsorption of L-cysteine on an aluminum surface followed the Langmuir adsorption isotherm and the polarization curves showed that the L-cysteine acts as a cathodic inhibitor to inhibit the cathodic reaction.

The adsorption energy was calculated on the surface of Al(1 1 1) and the L-cysteine molecules found negative values. The adsorption energy decreases in the order, (e) < (a) < (d) < (c) < (b) indicates that inhibitor molecules adsorbs on the surface of aluminum instead of water molecules and the adsorption is mainly based on the reactive groups (**Figure 4**). The interaction between inhibitor molecules and Al surface was further understood by calculations of charge density (**Figure 5**).

Marcus et al. reported corrosion inhibition of locally de-passivated surfaces by DFT study of 2-mercaptobenzothiazole on copper [163]. They have studied the adsorption of the organic inhibitor 2-mercaptobenzothiazole (MBT) on the incompletely passivated or locally depassivated copper surface based on quantum chemical DFT calculations. They investigated quantum chemical DFT calculations on the adsorption of MBT in thione or thiolate forms on a model of the copper surface incompletely covered by a surface oxide film, Cu(111) surface covered by an ultrathin Cu_2O oxide film (Cu(111)||Cu2O(111)). The calculation of DFT has been performed by periodic plane-wave basis set implemented in Vienna Ab initio Simulation Package (VASP) [70–73] with projector-augmented wave potentials using a 450 eV plane wave cutoff [74, 75]. Electron exchange and correlation terms were treated within the generalized gradient approximation (GGA) of Perdew–Burke–Ernzehof (PBE) functional [76, 77]. They used a Methfessel–Paxton smearing [78] with smearing value of 0.1 eV. They started from Cu (111)||Cu_2O (111) model followed by the construction of the hole in the oxide consists in removing certain copper and oxygen atoms from the surface and interface oxide layers. The hole in the oxide has been generated by removing eight atoms of copper and two atoms of oxygen, which correspond to four copper and one oxygen atom in each oxide layer. This study suggests that the partially oxidized copper surface blocks the initiation of pitting corrosion by strong interaction with the MBT molecule.

Allah and Moustafa studied quantum chemical calculations for amino pyrazole derivatives as inhibitors of corrosion of zinc, copper, and α-brass in an aqueous acid chloride solution [164]. They have performed the quantum mechanical calculations using the Dewar LCAO-SCF MO semi-empirical method, MNDO (modified neglect of differential overlap). The inhibition efficiency of the amino pyrazole derivatives depends upon the position of the substituent group with respect to the pyrazole ring, that is, the para, meta, or ortho positions of the arylazo group.

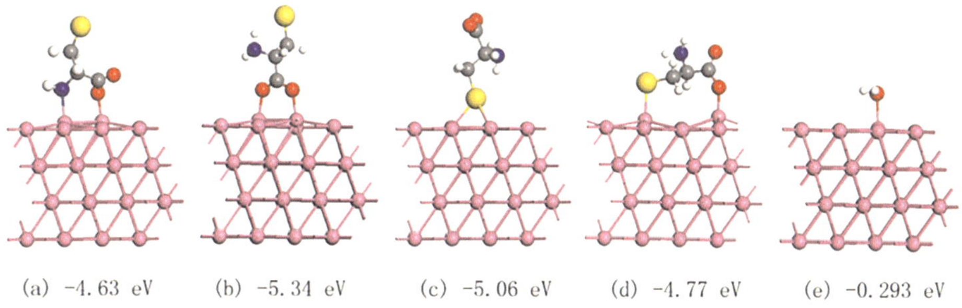

Figure 4.
Optimized structure and adsorption energy modes of L-cysteine molecules on Al (1 1 1).

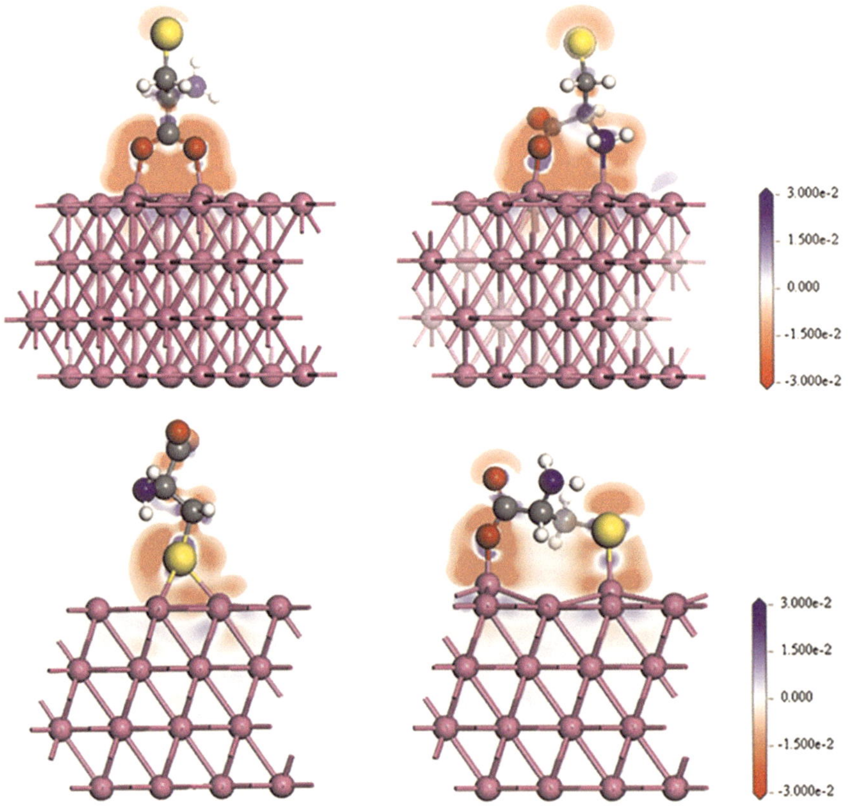

Figure 5.
Charge densities difference of different adsorption modes of L-cysteine molecules on Al (1 1 1).

Zhang et al. reported QSAR study on N-containing corrosion inhibitors: Quantum chemical approach assisted by the topological index [165]. They have studied the inhibition efficiency used DFTB3LYP/6-31G* level using GAUSSIAN 98 program package to carried out E_{homo}, E_{lumo}, energy gap, charge distribution, Dipole, topological index which refers to the steric hindrance of molecules.

Khadom has studied quantum chemical calculations of some amines corrosion inhibitors/copper alloy interaction in hydrochloric acid [166]. The quantum chemical calculations have been performed by Argus Lab 4.0.1 package, estimation by PM3 and AM1 method. The structural parameters, such as the frontier molecular orbital (MO), HOMO (highest occupied molecular orbital), LUMO (lowest unoccupied molecular orbital), dipole moment (μ), and ΔN were calculated. The inhibition efficiency calculation performed by AM1 and PM3 are in agreement with the experimental results.

Liang and Gao reported electrochemical and DFT studies of β-amino-alcohols (1-diethylamino-propan-2-ol (EAP) and 1,3-bis-diethylamino-propan-2-ol (DEAP)) as corrosion inhibitors for brass [167]. The DFT calculation was carried out with the Gaussian 98 program, B3LYP, and the 6-311G** orbital basis sets. The interactions of the inhibitor molecules with the metal surface were studied by the frontier orbital energy such as E_{HOMO}, E_{LUMO}, and energy difference (ΔE) and dipole moment (μ). The E_{HOMO} of DEAP (−0.205 a.u.) and EAP (−0.222 a.u.) and ΔE for EAP and DEAP are −0.258 and − 0.242 a.u., respectively. The dipole moment (μ) for DEAP and EAP are 2.32 and EAP 2.95, respectively. The theoretical calculations carried out for EAP and DEAP found that the inhibition efficiency of DEAP is higher than that of EAP which was in agreement with e experimental results potentiodynamic curves and electrochemical impedance spectroscopy (EIS).

Elhenawy and co-workers studied the understanding the adsorption performance of two glycine derivatives (bicine (N,N-bis(2-hydroxyethyl)glycine) and tricine (N-(tri(hydroxymethyl)methyl) glycine) as novel and environmentally safe anti-corrosion agents for copper in chloride solutions: experimental, DFT, and MC studies [168]. The quantum chemical indices (QCIs) for bicine, tricine, and glycine were performed using the density functional theory (DFT) with DMol3 module in Materials Studio software (Accelrys Inc.), the generalized gradient approximation (GGA) of the BOP function was used through the double-numeric basis set (DNP 3.5).

The Monte Carlo (MC) simulations studies were performed on Cu (111) surface, COMPASS force field was done for minimum energy calculation. The energies were calculated from the geometry optimized structure (with the energy minima), Mulliken charges were computed at the GGA/DNP 4.4 level for the atoms in bicine, tricine, and glycine molecules. Tricine was found higher inhibition efficiency than bicine or glycine as tricine DE values (2.9194, 3.0000, and 0.5920) and lower than those of bicine (3.9193, 3.5117, and 1.0343) and glycine (4.6342, 4.1916, and 1.1731) in the gas, solvated, and protonated molecules, respectively. Also, the ΔN values found from theoretical data are the order tricine > bicine > glycine. The adsorption energy against the Cu (111) surface for tricine was found more than that for bicine from Monte Carlo (MC) simulations. The experimental results were in agreement with computational indicating higher efficiency for tricine over that for bicine.

5. Heterocyclic compounds as organic corrosion inhibitor

Heterocyclic compounds containing N, S, O, and P atoms, as well as compounds containing π-bonds and/or polar group that will provide protection of metals or alloys from corrosion. The heterocyclic compounds such as azoles, indoles, and aromatic rings can be used to provide corrosion protection.

The benzo imidazole [169] derivative, pyridine thiazole [170] compound, imidazole derivatives [171], phenanthroimidazole derivatives [172], and bis-benzothiazoles derivatives [173] can be used as corrosion inhibitor potency for mild steel and others. Pyridine, Indoles, [174–177], and Quinoline [44, 178–181] have shown sufficient potency for inhibition of iron, aluminum, carbon steel, N80 steel, mild steel, and zinc in different acidic media.

6. Conclusions

In this chapter, we discussed the theoretical and molecular modeling of organic compounds as corrosion inhibitors. The inhibition behavior of organic compounds was analyzed by density functional theory (DFT), Molecular dynamic simulation (MD), Monte Carlo (MC) simulations, quantitative structure-activity relationship (QSAR) modeling, etc. The computational chemistry studies of organic compounds as corrosion inhibitor save time and money. This approach is environmental friendly as there is no need for synthesis of organic compounds.

Author details

Bharat Chandra Sahu
Department of Chemistry, Vishwavidyalaya Engineering College, Ambikapur,
A Constituent College of CSVTU, Bhilai, Chhattisgarh, India

*Address all correspondence to: bharatsahuiitb@gmail.com

References

[1] Bo¨hni H, Suter T, Schreyer A. Micro- and nanotechniques to study localized corrosion. Electrochimica Acta. 1995;**40**: 1361-1368

[2] Singh R, Dahotre NB. Journal of Materials Science: Materials in Medicine. 2007;**18**:725-751

[3] Davis JR. Corrosion: Understanding the basics. Materials Park, Ohio: ASM International; 2000

[4] Vargel C. Corrosion of Aluminium. United Kingdom: Elsevier; 2020

[5] Sadrabadi TE, Arjang M. A successful modification from corrosion engineering to corrosion management. In: 7th International Conference on Material Engineering and Metallurgy. Tehran-Iran; 2018

[6] Torsner E. Solving corrosion problems in biofuels industry. Corrosion Engineering, Science and Technology. 2010;**45**:42-48

[7] Van Delinder LS, deS Brasunas A. Corrosion Basics: An Introduction. Houston, TX: NACE; 1984

[8] Corrosion of Metals And Alloys – Basic Terms and Definitions. Brussels: International Organization for Standardization ISO 8044:1999; 2000

[9] Veronika KB. Knowledge about metals in the first century. Korrozios Figyelo. 2008;**48**:133-137

[10] Fontana MG. Corrosion Engineering. 3rd ed. New York: McGraw-Hill; 1986

[11] Verma C, Ebenso EE, Quraishi M. Journal of Molecular Liquids. 2017;**233**: 403-414

[12] Goyal M, Kumar S, Bahadur I, Verma C, Ebenso EE. Journal of Molecular Liquids. 2018;**256**:565-573

[13] Bennet LH, Kruger J, Parker RI, et al. Economic Effects of Metallic Corrosion in the United States. Washington, DC: National Bureau of Standards (Special Publication); 1978

[14] Koch GH, Brongers MPH, Thompson NG, et al. Materials Performance. 2002;**42**:3-11

[15] Kadhim A, Al-Amiery AA, Alazawi R, Al-Ghezi MKS, Abass RH. International Journal of Corrosion and Scale Inhibition. 2021;**10**:54-67

[16] Raja PB, Ismail M, Ghoreishiamiri S, Mirza J, Ismail MC, Kakooei S, et al. Chemical Engineering Communications. 2016;**203**:1145-1156

[17] Popoola LT. Corrosion Reviews. 2019;**37**:71-102

[18] Verma C, Verma DK, Ebenso EE, Quraishi MA. Sulfur and phosphorus heteroatom-containing compounds as corrosion inhibitors. Heteroatom Chemistry. 2018;**19**:1-20

[19] Robertson WD. Journal of the Electrochemical Society. 1951;**98**:94-100

[20] Pryor MJ, Cohen M. Journal of the Electrochemical Society. 1953;**100**: 203-215

[21] Al-Borno A, Islam M, Khraishi R. Corrosion. 1989;**45**:970-975

[22] Alexander DB, Moccari AA. Corrosion. 1993;**49**:921-928

[23] Yashiro H, Oyama A, Tanno K. Corrosion. 1997;**53**:290-297

[24] Danaee I, Niknejad Khomami M, Attar AA. Journal of Materials Science and Technology. 2013;**29**:89-96

[25] Carrillo I, Valdez B, Zlatev R, Stoytcheva M, Carrillo M, Bäbler R. International Journal of Electrochemical Science. 2012;**7**:8688-8701

[26] Shams El Din AM, Wang L. Desalination. 1996;**107**:29-43

[27] Samiento-Bustos E, González Rodriguez JG, Uruchurtu J, Dominguez Patiño G, Salinas-Bravo VM. Corrosion Science. 2008;**50**:2296-2303

[28] Zhao JM, Zuo Y. Corrosion Science. 2002;**44**:2119-2130

[29] Ilevbare GO, Burstein GT. Corrosion Science. 2003;**45**:1545-1569

[30] Foad El-Sherbini EE. Corrosion Science. 2006;**48**:1093-1105

[31] Foad El-Sherbini EE, Abd-El-Wahab SM, Amin MA, Deyab MA. Electrochemical Corrosion Science. 2006;**48**:1885-1898

[32] Abdel-Gaber AM, Abd-El-Nabey BA, Khamis E, Abd-El-Khalek DE. Desalination. 2011;**278**:337-342

[33] Blustein G, Romagnoli R, Jaén JA, Di Sarli AR, del Amo B. Colloids and Surfaces A: Physicochemical and Engineering Aspects. 2006;**290**:7-18

[34] Bommersbach P, Alemany-Dumont C, Millet JP, Normand B. Electrochimica Acta. 2005;**51**:1076-1084

[35] Lecante A, Robert F, Blandinières PA, Roos C. Current Applied Physics. 2011;**11**:714-724

[36] Verma C, Verma DK, Ebenso EE, Quraishi MA. Heteroatom Chemistry. 2018;**29**:1-20

[37] Verma DK, Dewangan Y, Dewangan AK. Journal of Bio- and Tribo-Corrosion. 2021;**7**:15

[38] Radojčić I, Berković K, Kovač S, Vorkapić-Furač J. Corrosion Science. 2008;**50**:1498-1504

[39] Mishra A, Aslam J, Verma C, Quraishi M, Ebenso EE. Journal of the Taiwan Institute of Chemical Engineers. 2020;**114**:341-358

[40] Lebrini M, Robert F, Vezin H, Roos C. Corrosion Science. 2010;**52**: 3367-3376

[41] Moretti G, Quartarone G, Tassan A, Zlngales A. Electrochimica Acta. 1996; **41**:1971-1980

[42] Abd El-Maksoud S, Fouda A. Materials Chemistry and Physics. 2005; **93**:84-90

[43] Ogretir C, Mihci B, Bereket G. Journal of Molecular Structure: THEOCHEM. 1999;**488**:223-231

[44] Verma C, Quraishi M, Ebenso EE. Surfaces and Interfaces. 2020;**21**:100634

[45] Lavanya K, Saranya J, Chitra S. Recent reviews on quinoline derivatives as corrosion inhibitors. Corrosion Reviews. 2018;**36**:365-371

[46] Prajapati SM, Patel KD, Vekariya RH, Panchal SN, Patel HD. RSC Advances. 2014;**4**:24463-24476

[47] Zarrouk A, Hammouti B, Lakhlifi T, Traisnel M, Vezin H, Bentiss F. Electrochemical, XPS and DFT studies. Corrosion Science. 2015;**90**: 572-584

[48] Ibrahim I, Yunus S, Hashim M. International Journal of Scientific and Engineering Research. 2013;**4**:1-12

[49] Choi H, Kim KY, Park JM. Progress in Organic Coatings. 2013;**76**:1316-1324

[50] Fouda A, Mostafa H, El-Taib F, Elewady G. Corrosion Science. 2005;**47**: 1988-2004

[51] Stupnisek-Lisac E, Brnada A, Mance AD. Corrosion Science. 2000;**42**: 243-257

[52] Desai M, Shah V. Aromatic Corrosion Science. 1972;**12**:725-730

[53] Kuznetsov YI. Organic Inhibitors of Corrosion of Metals. New York: Springer Science; 1996

[54] Quraishi MA, Chauhan DS, Saji VS. Heterocyclic Organic Corrosion Inhibitors Principles and Applications. United Kingdom: Elsevier; 2020

[55] Verma C, Hussain CM, Ebenso EE, editors. Organic Corrosion Inhibitors Synthesis, Characterization, Mechanism, and Applications. United States: Wiley; 2022

[56] Yang H-M. Molecules. 2021;**26**:3473

[57] Gupta NK, Quraishi MA, Verma C, Mukherjee AK. RSC Advances. 2016;**6**: 102076-102087

[58] Verma C, Haquec J, Quraishi MA, Ebenso EE. Journal of Molecular Liquids. 2019;**275**:18-40

[59] Verma C, Ebenso EE, Quraishi MA, Hussain CM. Materials Advances Review. 2021;**2**:3806-3850

[60] Verma CB, Quraishi M, Singh A. Journal of the Taiwan Institute of Chemical Engineers. 2015; **49**:229-239

[61] Solmaz R, Altunbaş E, Kardaş G. Materials Chemistry and Physics. 2011; **125**:796-801

[62] Solmaz R. Corrosion Science. 2010; **52**:3321-3330

[63] El-Haddad MN. International Journal of Biological Macromolecules. 2013;**55**:142-149

[64] Hosseini M, Mertens SF, Arshadi MR. Corrosion Science. 2003;**45**: 1473-1489

[65] Obot I, Obi-Egbedi N, Umoren S. Corrosion Science. 2009;**51**:1868-1875

[66] Li W, He Q, Pei C, Hou B. Electrochimica Acta. 2007;**52**: 6386-6394

[67] Lozano I, Mazario E, Olivares-Xometl C, Likhanova N, Herrasti P. Materials Chemistry and Physics. 2014; **147**:191-197

[68] Awad MK, Mustafa MR, Elnga MMA. Journal of Molecular Structure: THEOCHEM. 2010;**959**:66-74

[69] Zhang Q, Gao Z, Xu F, Zou X. Colloids and Surfaces A: Physicochemical and Engineering Aspects. 2011;**380**:191-200

[70] Dussan E. On the spreading of liquids on solid surfaces: Static and dynamic contact lines. Annual Review of Fluid Mechanics. 1979;**11**:371-400

[71] Aveyard R, Haydon DA. An Introduction to the Principles of Surface Chemistry. Cambridge, England: Cambridge University Press; 1973

[72] Likhanova NV, Domínguez-Aguilar MA, Olivares-Xometl O, Nava-Entzana N, Arce E, Dorantes H. Corrosion Science. 2010;**52**:2088-2097

[73] de Chialvo MG, Chialvo A. Electrochimica Acta. 1998;**44**:841-851

[74] de Chialvo MG, Chialvo A. Electrochemistry Communications. 1999;**1**:379-382

[75] Yin ZF, Feng YR, Zhao WZ, Bai ZQ, Lin GF. Surface and Interface Analysis. 2009;**41**:517-523

[76] Xiang Y, Wang Z, Li Z, Ni WD. Corrosion Engineering, Science and Technology. 2013;**48**:121-129

[77] Ebenso EE, Ekpe UJ, Ita BI, Offiong OE, Ibok UJ. Materials Chemistry and Physics. 1999;**60**:79-90

[78] Muralidharan S, Quraishi M, Iyer S. Corrosion Science. 1995;**37**: 1739-1750

[79] Verma C, Olasunkanmi L, Ebenso EE, Quraishi M. Journal of Molecular Liquids. 2017;**251**:100-118

[80] Leimkuhler B, Matthews C. Molecular Dynamics. Switzerland: Springer; 2015

[81] Hollingsworth SA, Dror RO. Neuron. 2018;**99**:1129-1143

[82] Pareek S, Jain D, Hussain S, Biswas A, Shrivastava R, Parida SK, et al. Chemical Engineering Journal. 2019;**358**: 725-742

[83] Zhang XY, Kang QX, Wang Y. Computational and Theoretical Chemistry. 2018;**1131**:25-32

[84] Ebenso EE, Verma C, Olasunkanmi LO, Akpan ED, Verma DK, Lgaz H, et al. Physical Chemistry Chemical Physics. 2021;**23**: 19987-20027

[85] Verma C, Lgaz H, Verma DK, Ebenso EE, Bahadur I, Quraishi MA. Journal of Molecular Liquids. 2018;**260**: 99-120

[86] Verma DK, Khan F, Bahadur I, Salmane M, Quraishie MA, Verma C, et al. Results in Physics. 2018; **10**:665-674

[87] Vermaa DK, Ebensob EE, Quraishi MA, Verma C. Results in Physics. 2019;**13**:102194

[88] Verma DK, Fantazi AA, Verma C, Khan F, Asatkar A, Hussain CM, et al. Journal of Molecular Liquids. 2020;**314**: 113651

[89] Verma DK, Kaya S, E-chihbi E, El-Hajjaji F, Phukan MM, Alnashiri HM. Journal of Molecular Liquids. 2021;**329**: 115531

[90] Verma DK, Aslam R, Aslam J, Quraishi MA, Ebenso EE, Verma C. Journal of Molecular Structure. 2021; **1236**:130294

[91] Verma DK, Kazi M, Alqahtani MS, Syedb R, Berdimurodov E, Kaya S, et al. Journal of Molecular Structure. 2021; **1241**:130648

[92] Oguzie EE, Li Y, Wang SG, Wang F. RSC Advances. 2011;**1**:866-873

[93] Verma C, Quraishi MA, Ebenso EE, Obot IB, Assyry AE. Journal of Molecular Liquids;**219**(2016):647-660

[94] Mishraa A, Vermab C, Lgaz H, Srivastavaa V, Quraishia MA, Ebenso EE. Journal of Molecular Liquids. 2018;**251**:317-332

[95] Mishra A, Verma C, Srivastava V, Lgaz H, Quraishi MA, Ebenso EE, et al. Journal of Bio- and Tribo-Corrosion. 2018;**4**:32

[96] Taylor CD. Modelling corrosion, atom by atom. The Electrochemical Society Interface. 2014;**23**:59-64

[97] Al-Mobarak NA, Khaled KF, Hamed MNH, Abdel-Azim KM, Abdelshafi NS. Arabian Journal of Chemistry. 2010;**3**:233-242

[98] Liu J, Yu W, Zhang J, Hu S, You L, Qiao G. Applied Surface Science. 2010; **256**:4729-4733

[99] Khaled KF. Material Chemical and Physics. 2010;**124**:760-767

[100] Khaled KF. Journal of Applied Electrochemistry. 2011;**41**: 423-433

[101] Khaled KF, Abdelshafi NS, El-Maghraby A, Al-Mobarak N. Journal of Materials and Environmental Science. 2011;**2**:166-173

[102] Khaled KF, Abdel-Shafi NS. International Journal of Electrochemical Science. 2011;**6**:4077-4094

[103] Khaled KF. Journal of Chemica Acta. 2012;**1**:66-71

[104] Khaled KF, Abdelshafi NS, El-Maghraby AA, Aouniti A, Al-Mobarak N, Hammout B. International Journal of Electrochemical Science. 2012;**7**:12706-12719

[105] Khaled KF, Amin MA. Journal of Applied Electrochemistry. 2008;**38**: 1609-1621

[106] Khaled KF. International Journal of Electrochemical Science. 2012;**7**: 1027-1044

[107] Obot IB, Obi-Egbedi NO, Ebenso EE, Afolabi AS, Oguzie EE. Research on Chemical Intermediates. 2013;**39**:1927-1948

[108] Umoren SA, Obot IB, Gasem ZM. Ionics. 2015;**21**:1171-1186

[109] Singh P, Ebenso EE, Olasunkanmi LO, Obot IB, Quraishi MA. The Journal of Physical Chemistry C. 2016;**120**:3408-3419

[110] Guo L, Ye G, Obot IB, Li X, Shen X, Shi W, et al. International Journal of Electrochemical Science. 2017;**12**:166-177

[111] Abdelahi MMM, Elmsellem H, Benchidmi M, Sebbar NK, Belghiti MA, Ouasif LEI, et al. Journal of Materials and Environmental Science. 2017;**8**: 1860-1876

[112] Sikine M, Elmsellem H, Rodi YK, Kadmi Y, Belghiti M, Steli H, et al. Experimental, Journal of Materials and Environmental. Science. 2017;**8**:116-133

[113] Xia S, Qiu M, Yu L, Liu F, Zhao H. Corrosion Science. 2008;**50**:2021-2029

[114] Khaled KF. Electrochimica Acta. 2008;**53**:3484-3492

[115] Khaled KF, Amin MA. Journal of Applied Electrochemistry. 2009;**39**: 2553-2568

[116] Khaled KF. Journal of Solid State Electrochemistry. 2009;**13**:1743-1756

[117] Khaled KF, Amin MA. Corrosion Science. 2009;**51**:1964-1975

[118] Khaled KF, Fadl-Allah SA, Hammouti B. Materials Chemistry and Physics. 2009;**117**:148-155

[119] Obot IB, Haruna K, Saleh TA. Arabian Journal for Science and Engineering. 2019;**44**:1-32

[120] Marcus P, Maurice V. Philosophical Transactions of the Royal Society A. 2017;**375**:1-14

[121] Ke H, Taylor CD. Corrosion. 2019; **75**:708-726. DOI: 10.5006/3050

[122] Costa D, Ribeiro T, Cornette P, Marcus P. The Journal of Physical Chemistry C. 2016;**120**:28607-28616

[123] Obot IB, Macdonald DD, Gasem ZM. Corrosion Science. 2015;**99**: 1-30

[124] Obot IB, Kaya S, Kaya C, Tüzün B. Physica E: Low-dimensional Systems and Nanostructures. 2016;**80**:82-90

[125] Wazzan NA, Mahgoub FM. Open Journal of Physical Chemistry. 2014;**4**: 6-14

[126] Koopmans T. Über die Zuordnung von Wellenfunktionen und Eigenwerten zu den einzelnen Elektronen eines Atoms. Physica. 1934;**1**:104-113

[127] Parr RG, Pearson RG. Journal of the American Chemical Society. 1983;**105**: 7512-7516

[128] Lukovits I, Kalman E, Zucchi F. Corrosion. 2001;**57**:3-8

[129] Lgaz H, Chung IM, Albayati MR, et al. Arabian Journal of Chemistry. 2020;**13**:2934-2954

[130] Lgaz H, Salghi R, Masroor S, Kim S-H, Kwon C, Kim SY, et al. Journal of Molecular Liquids. 2020;**308**: 112998

[131] Perdew JP, Burke K, Ernzerhof M. Physical Review Letters. 1996;**77**:3865

[132] Štich I. Physical Review A. 1991;**44**: 1401-1404

[133] McCammon JA, Gelin BR, Karplus M. Dynamics of folded proteins. Nature. 1977;**267**:585-590

[134] Stillinger FH, Rahman A. The Journal of Chemical Physics. 1974;**60**: 1545-1557

[135] Alder BJ, Wainwright TE. The Journal of Chemical Physics. 1957;**27**: 1208-1209

[136] Ojeda P, Garcia ME, Londoño A, Chen N-Y. Biophysical Journal. 2009;**96**: 1076-1082

[137] Milik M, Skolnick J. Proteins: Structure, Function, and Genetics. 1993; **15**:10-25

[138] Meller JA. Molecular dynamics. In: Encyclopedia of Life Sciences. Nature Publishing Group. 2001;1-8

[139] Hansson T, Oostenbrink C, van W. Gunsteren. Current Opinion in Structural Biology. 2002;**12**:190–196

[140] Theodorou DN. Diffusion in Polymers. New York: Marcel Dekker; 1996

[141] Casewit CJ, Colwell KS, Rappe AK. Journal of the American Chemical Society. 1992;**114**:10046-10053

[142] Rappe AK, Casewit CJ, Colwell KS, et al. Journal of the American Chemical Society. 1992;**114**:10024-10035

[143] Hagler AT, Huler E, Lifson S. Journal of the American Chemical Society. 1974;**96**:5319-5327

[144] Hagler AT, Lifson S. Journal of the American Chemical Society. 1974;**96**: 5327-5335

[145] Lifson S, Hagler AT, Dauber P. Journal of the American Chemical Society. 1979;**101**:5111-5121

[146] Rapaport DC. The Art of Molecular Dynamics Simulation. Cambridge, England: Cambridge University Press; 2004

[147] Taylor CD, Marcus P. Molecular Modeling of Corrosion Processes. Hoboken: Wiley; 2015

[148] Saha SK, Dutta A, Ghosh P, et al. Physical Chemistry Chemical Physics. 2016;**18**:17898-17911

[149] Lgaz H, Bhat KS, Salghi R, Shubhalaxmi, Jodeh S, Algarrae M, et al. Journal of Molecular Liquids. 2017;**238**: 71-83

[150] Lgaz H, Salghi R, Bhat KS, Chaouikia A, Shubhalaxmi, Jodeh S. Journal of Molecular Liquids. 2017;**244**: 154-168

[151] Chugh B, Singh AK, Thakur S, Thakur S, Pani B, Pandey AK, et al. Journal of Physical Chemistry C. 2019; **123**:22897-22917

[152] Lgaz H, Chung IM, Salghi R, Ali IH, Chaouiki A, Aoufir Y, et al. Applied Surface Science. 2019;**463**:647-658

[153] Chaouiki A, Chafiq M, Ko YK, Al-Moubaraki AH, Thari, FZ, Salgi R, et al.. Metals. 2022;**12**:1598

[154] Fu J, Li S, Wang Y, Cao L, Lu L. Journal of Materials Science. 2010;**45**: 6255-6265

[155] Amin MA, Khaled KF, Mohsen Q, Arida HA. Corrosion Science. 2010;**52**: 1684-1695

[156] Obot IB, Gasem ZM, Umoren SA. International Journal of Electrochemical Science. 2014;**9**:2367-2378

[157] Chan-Rosado G, Pech-Canul MA. International Journal of Corrosion. 2016:1-12

[158] Gupta S, Gupta KK, Andersson M, Yazdi R, Ambat R. Corrosion Science. 2022;**195**:109999

[159] Sahaa SK, Murmua M, Murmua NC, Obot IB, Banerjee P. Surfaces and Interfaces. 2018;**10**:65-73

[160] Zor S, Kandemirli F, Bingul M. Protection of Metals and Physical Chemistry of Surfaces. 2009;**45**:46-53

[161] Kumar H, Kumari M. Journal of Molecular Structure. 2021;**1229**:129598

[162] Wang D, Gao L, Yang D, Wang H, Lin T. Materials Chemistry and Physics. 2016;**169**:142-151

[163] Chiter F, Costa D, Maurice V, Marcus P. NPJ Materials Degradation. 2021;**52**:1-12

[164] Gad Allah AG, Moustafa H. Journal of Applied Electrochemistry. 1992;**22**: 644-648

[165] Zhang SG, Lei W, Xia MZ, Wang FY. Journal of Molecular Structure: THEOCHEM. 2005;**732**:173-182

[166] Khadom AA. Journal of Materials and Environmental Sciences. 2017;**8**: 1153-1160

[167] Gao G, Liang C. Electrochim Acta. 2007;**52**:4554-4559

[168] Elgendy A, Nady H, El-Rabieib MM, Elhenawy AA. RSC Advances. 2019;**9**:42120-42131

[169] Srivastava V, Salman M, Chauhan DS, et al. Journal of Molecular Liquids. 2021;**324**:115010

[170] Farahati R, Behzadi H, Mousavi-Khoshdel SM, Ghaffarinejad A. Journal of Molecular Structure. 2020;**1205**:127658

[171] Ouakki M, Galai M, Rbaa M, et al. Journal of Molecular Liquids. 2020;**319**: 114063

[172] Lv Y-L, Kong F-Y, Zhou L, et al. Journal of Molecular Structure. 2021; **1228**:129746

[173] Suhasaria A, Murmu M, Satpati S, et al. Journal of Molecular Liquids. 2020; **313**:113537

[174] Ituen E, Mkpenie V, Moses E, Obot I. Bioelectrochemistry. 2019;**129**: 42-53

[175] Abdellaouia O, Skallia MK, Haoudia A, et al. Chem. 2021;**9**:44-56

[176] Keleşoglu A, Yıldız R, Dehri I. Journal of Adhesion Science and Technology. 2019;**33**:2010-2030

[177] Feng L, Zhang S, Feng Y, et al. Chemical Engineering Journal. 2020; **394**:124909

[178] Fernandes CM, Faro LV, Pina VGSS, et al. Surfaces and Interfaces. 2020;**21**:100773

[179] Douche D, Elmsellem H, Anouar EH, et al. Journal of Molecular Liquids. 2020;**308**:113042

[180] Rbaaa M, Ouakkib M, Galaic M, et al. Colloids and Surfaces, A: Physicochemical and Engineering Aspects. 2020;**602**:125094

[181] Erazua EA, Radeleke BB. Journal of Applied Sciences and Environmental Management. 2019;**23**:1819-1824

Section 2

Coatings Used in Corrosion

Chapter 5

Self-Healing Coatings

Shahin Kharaji

Abstract

Metal corrosion is a natural and inevitable process that imposes a lot of cost on many industries and can also have irreparable consequences. Several methods, such as cathodic protection, galvanizing, painting, and coatings, are available to prevent metal corrosion. Selection of the best corrosion prevention method depends on many factors including cost, effectiveness, type of metal, and corrosive media but it can be said that coatings are probably the most convenient method to prevent corrosion of metals due to the low cost, availability of raw materials, flexibility, and simplicity. Despite having many advantages, coatings are subject to problems such as cracking and degradation. Therefore, they must be repaired or replaced. Self-healing coating has been introduced and developed during the past decades as a very effective method to overcome the problems of traditional coatings. Self-healing means healing (recover/repair) internal damages automatically and autonomously. It is an amazing property that can fill cracks and small pinholes which leads to increased service lives of coatings. This chapter presents different strategies for fabrication of self-healing materials and explains their challenges and limitations. Furthermore, the use of self-healing materials in metal corrosion through different mechanisms is discussed, and published reports in this field are reviewed.

Keywords: metal corrosion, coating, self-healing materials, autonomous healing mechanisms, non-autonomous healing mechanisms

1. Introduction

Corrosion can be defined as an attack on materials because of an interaction of materials with the surrounding environment. Based on this definition, all materials can corrode. But in practice, the term corrosion is mainly associated with the degradation of metals [1]. Metals are widely used materials due to having remarkable properties such as high electrical and thermal conductivity, ductility, malleability, and high strength. Apart from gold, platinum, and a few others, most metals are prone to corrosion. This problem can be more important when using metal alloys as the type, speed, and rate of corrosion are different for each element. In fact, more than one type of corrosion can occur in the corrosion of alloys. Corrosion is a common and costly phenomenon that should be prevented to avoid irreparable risks such as damage to structures and human life. There are several factors that influence the rate of corrosion including diffusion, temperature, conductivity, humidity, type of ions, pH value, and electrochemical potential [2]. The corrosion rate of metals can be controlled or reduced by taking various measurements. The most common methods include cathodic protection, galvanizing, painting, and coatings. Coatings are probably the most convenient method to prevent the corrosion of metals due to the low cost, availability of raw materials, flexibility, and simplicity. There are many different types of industrial coatings, including organic (alkyd coatings, epoxy coating, polyurethane coatings, etc.) and inorganic coatings (ceramic coatings, metalized coating, etc.).

Despite having many advantages, coatings are subject to problems such as cracking and degradation. Therefore, they must be repaired or replaced. Being exposed to a corrosive environment for a long time leads to the weakening of the mechanical properties of the coating, and as a result, microcracks form and propagate. The failure of the coatings is inevitable and hence they must be repaired or replaced. In order to improve the properties of the coatings, several methods such as adding fillers in a polymer matrix have been proposed and implemented. Various fillers including zinc oxide, titanium dioxide, and silicon dioxide have been used and commercialized which enhance both mechanical properties and corrosion inhibition performance of coatings. During the past two decades, new-generation protective coatings, self-healing coatings, have been introduced and developed that can repair the damage caused by corrosive media and as a result, increase the coating lifetime. Self-healing coating has become a research hotspot in the coating field. Thus underlying their properties, mechanism, and synthesis procedures could be interesting. However, the development of these coatings is still in its early stages and it is very important to know their various aspects, including their properties, mechanism, performance, and application. This chapter provides basic information about metal corrosion and the development of methods to inhibit this phenomenon. In addition, different mechanisms of self-healing materials and published reports in this field are reviewed.

2. Basics of metal corrosion

Commonly, metal corrosion is defined as the deterioration of metals as a result of exposure to a corrosive environment, such as the atmosphere or even more aggressive corrosive media like seawater and chemical plants [3]. Definitely, our daily life is negatively affected by metal corrosion as automobiles, buildings, infrastructure, appliances, and energy distribution systems are all susceptible to corrosion because of metal pieces. Metals are rarely used in pure form (only in special applications), and they are generally used as alloys. There are many debates about which pure metal or alloy is the best, but what is clear is that alloys can provide unique properties. For instance, stainless steel, which is an alloy of iron, is more resistant to corrosion and has a slower corrosion rate than pure iron. In addition, it is an extremely tough and highly durable material with high impact resistance in comparison with pure iron [4]. But this does not necessarily mean that alloys do not corrode. Therefore, it can be said that corrosion in metals and even alloys is inevitable and measures must be taken to prevent it. Before pointing out the methods to prevent metal corrosion, it is beneficial to understand the causes and mechanisms of metal corrosion. Although corrosion is mainly a chemical phenomenon, mechanical factors, such as tensile or compressive forces and wear, can also aggravate it in many situations called mechanically assisted corrosion. For example, in offshore environments, waves, tides, and strong winds can cause mechanically assisted corrosions such as stress-corrosion cracking, fatigue corrosion, corrosion-erosion, and fretting-corrosion [5]. **Figure 1** depicts the most common types of mechanically assisted corrosion.

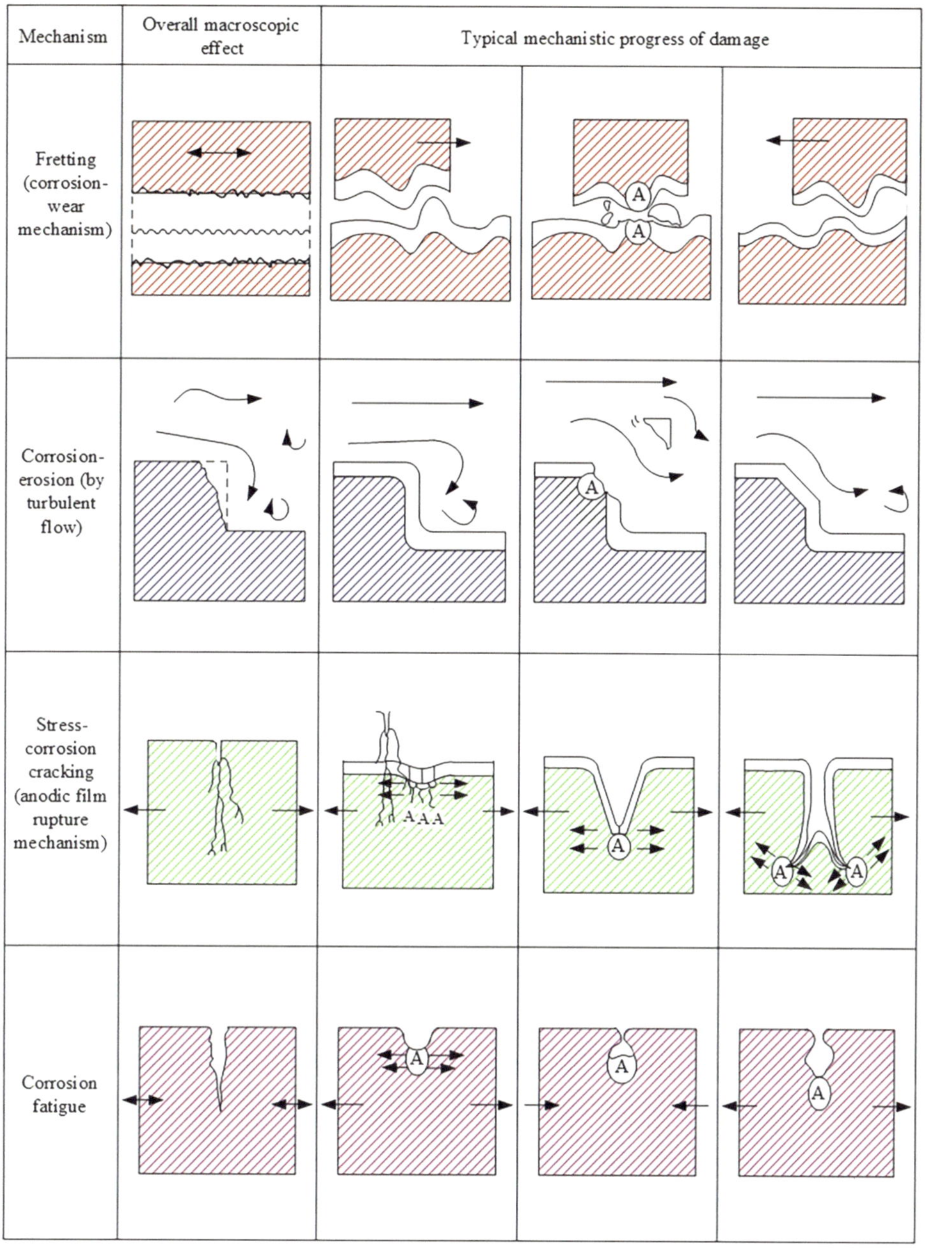

Figure 1.
The most probable types of mechanically assisted corrosion [5].

Generally, Corrosion consists of a series of complex chemical reactions and may initiate by several different mechanisms that depend on the surrounding environment and involves simultaneous oxidation and reduction reactions. In fact, first, an aqueous adlayer forms on the metal surface followed by electrochemical and chemical reactions. Furthermore, corrosion products may participate in these reactions. The mechanism and rate of chemical reactions that lead to corrosion depending on the corrosive factors in the surrounding environment which cause various classifications of corrosion. The most important factor contributing to metal corrosion can be classified as following [6]:

- Gases (CO_2, SO_x, HCL, HF, H_2SO_4, NH_3, H_2S, etc.)
- Moisture, dew, and condensation
- Temperature
- Relative humidity
- Type of metal
- Acidity and alkalinity (pH)
- Corrosive ions (Cl^-, SO_4^{2-}, Mg^{2+}, etc.)

The most common types of metal corrosion can be listed as follows:

- **Uniform corrosion**

Uniform corrosion is one of the most typical types of corrosion and occurs when the entire surface area is exposed to a corrosive environment. Uniform corrosion is one of the most typical types of metal corrosion and occurs when the entire surface area is exposed to a corrosive environment. In this type of corrosion, the thickness of the metal decreases over time, and if it is not prevented, it continues until the metal is completely destroyed [7]. **Figure 2** shows the uniform corrosion in metals.

- **Galvanic/bimetallic corrosion**

Galvanic corrosion takes place when dissimilar metals come in contact with each other when exposed to an electrolyte. In this type of corrosion, one metal deteriorates quickly by an electrochemical reaction while the other remains unaffected. This is

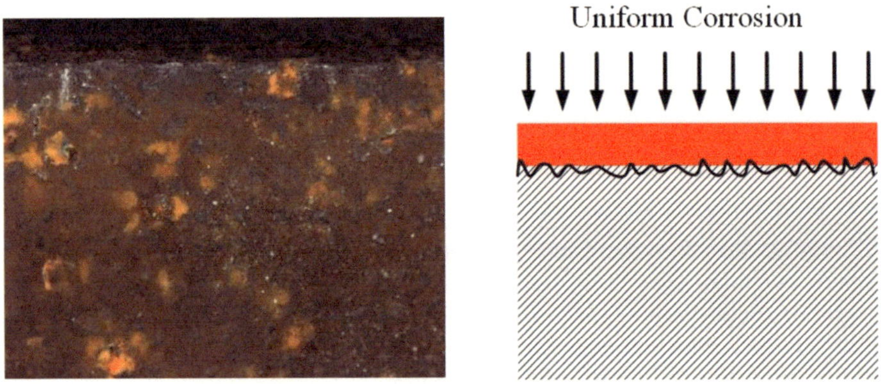

Figure 2.
The uniform corrosion on metal surface [8].

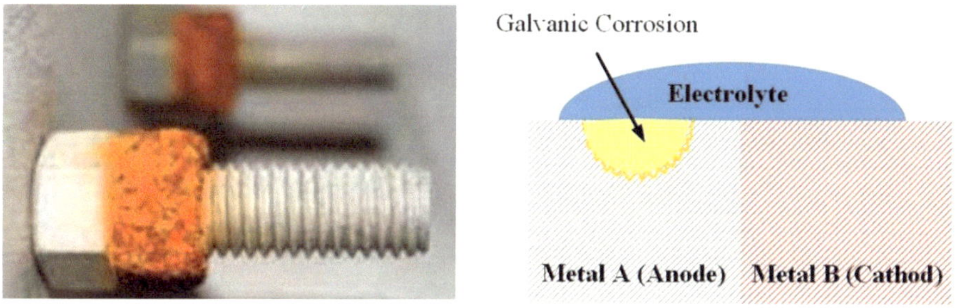

Figure 3.
The galvanic corrosion in metals [9].

due to the magnitude of the potential difference between the two metals [9]. **Figure 3** illustrates the galvanic corrosion in metals.

- **Crevice corrosion**

Crevice corrosion is a localized type of corrosion and is formed inside gaps or crevices on the surface of the metal. This type of corrosion can also affect the connected pieces to the metal substrate such as another welded metal or even metal attached to non-metal objects. The filling of crevices by corrosive agents, such as water, humidity, or pollutants, causes the rate of this type of corrosion to increase and eventually leads to the failure of the substrate [10]. **Figure 4** depicts the crevice corrosion in metals.

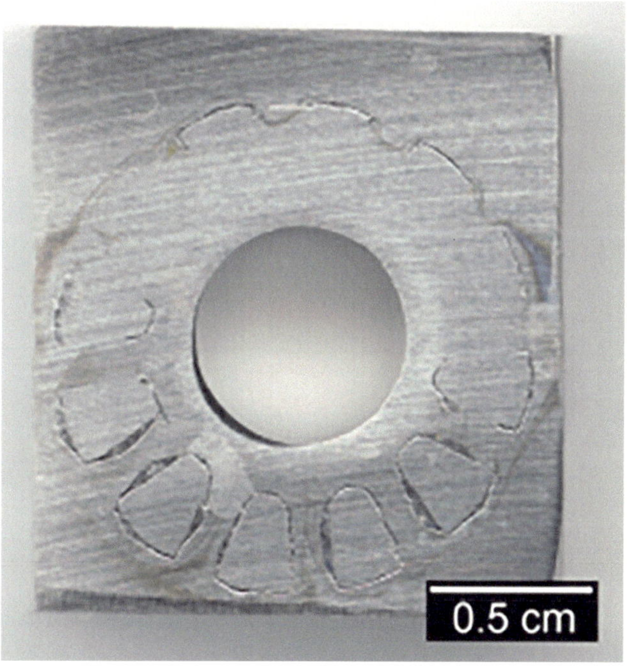

Figure 4.
The crevice corrosion in metals [10].

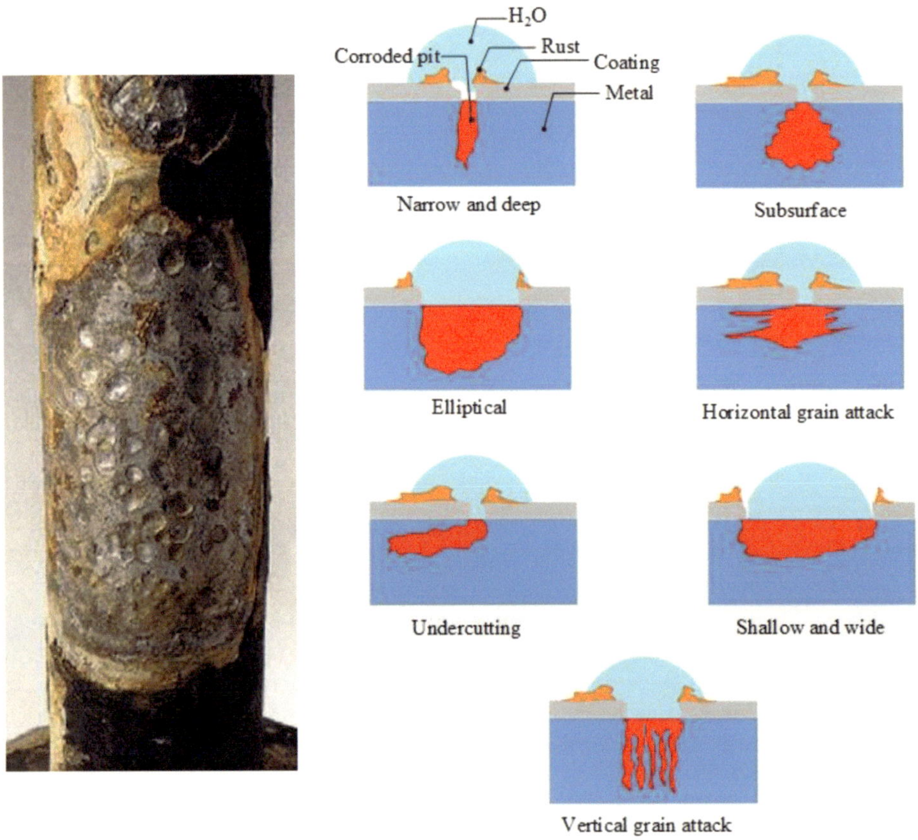

Figure 5.
The pitting corrosion in metals [11].

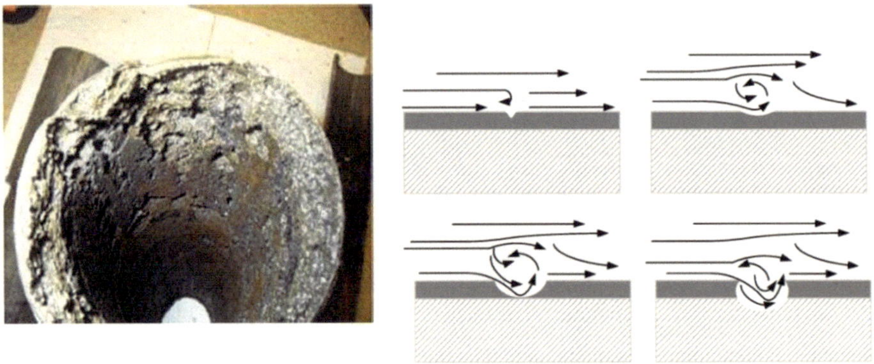

Figure 6.
The erosion-corrosion in metals [12].

- **Pitting corrosion**

Pitting corrosion is a localized form of corrosion that creates holes on the surface of metals. This type of corrosion takes place when a small area of metal becomes anodic while a vast area of the substrate acts as a cathode. The shape of holes may not be similar but mostly they are hemispherical and their size can vary from shallow and wide to deep and narrow. The formation and growth of holes on the surface of metals can have irreparable consequences, especially in critical points such as joint and welded parts [11]. **Figure 5** shows the pitting corrosion in metals.

- **Erosion-corrosion**

Erosion-corrosion is a form of corrosion that occurs as a result of chemical and mechanical stimuli. The mechanical effect of flow or velocity of a fluid combined with the corrosive action of the fluid causes accelerated loss of metal. This type of corrosion initiates by removing parts of the metal by erosion and is followed by contact with a corrosive medium. This usually happens when the velocity of the fluid is too high (turbulent flow) such as in tube blockages, tube inlet ends, or pump impellers. The process is a cyclic process that continues until the perforation of the metal occurs [12]. **Figure 6** illustrates the erosion-corrosion in metals.

3. Corrosion protection methods

As stated before, corrosion is one of the leading problems faced by almost all industries and can lead to safety issues and ruin the integrity of equipment and supplies. Therefore, steps should be taken to prevent or minimize metal corrosion. There are various methods to protect metals from corrosion such as hot dip galvanization, cathodic protection, painting, and coating. Choosing the best method depends on cost, effectiveness, type of metal, corrosive media, etc. Cathodic protection can perform in two different ways, the impressed current cathodic protection (ICCP) system and the sacrificial anodes cathodic protection (SACP) system. In the ICCP system, the metal substrate is connected to an external DC power source and the resulting current prevents the corrosion process. The SCAP is a technique that uses a less noble metal to act as the anode. In this method, an active metal such as magnesium, aluminum, and zinc is connected to the substrate that needs to be protected and provides electrons to the structure to be protected and consumed [13]. The ICCP systems require high initial costs due to the need for an external power supply as well as skilled workers, but it is more effective than SACO systems. [14].

In the hot-dip galvanization method, fabricated steel is immersed in molten zinc which leads to forming of a series of zinc-iron alloy layers on the metal surface by a metallurgical reaction between the iron and zinc. This method is very effective for protecting steel as zinc oxide (ZnO) is formed by the reaction of zinc and oxygen which provides a robust barrier to steel corrosion [15].

Painting is another way to protect metals from corrosion. Paints are a thin passive layer that typically contains a complex mix of ingredients, each with its own purpose. These can include binders that help glue the paint together, pigments that give the paint its color and other useful properties, solvents that help the ingredients mix properly, and many others. Paints are mostly used for decorative purposes that can protect the metal to some extent from exposure to corrosive agents. However, if the paint film is damaged, corrosion will occur very quickly due to a lack of protection from the environment [16].

Coating is one of the most convenient and effective methods for the prevention of metal corrosion. Coating is a widely used and effective method due to the availability of raw materials, simplicity, low cost, flexibility, and versatility [17]. Despite having many advantages, coatings face problems like other corrosion protection methods. During the service life, the mechanical properties of the coating film change which leads to the formation of cracks. Sometimes cracks are deep within the structure where detection is difficult and repair is almost impossible. This leads to the destruction of the coating and as a result, the substrate is exposed to a corrosive environment [18]. Self-healing materials have been introduced and developed in recent years as a novel method to overcome this drawback. Self-healing materials can repair minor damage automatically and autonomously. Self-healing materials have revolutionized the coating systems which lead to a significant increase in the service life of coatings [18]. Therefore, the following section covers different aspects of these novel materials including the definition of self-healing materials, design strategies, effective parameters, and their mechanisms.

4. Self-healing materials

To better understand the basis of self-healing materials, it would be better to define self-healing meaning first. Self-healing materials can heal (recover/repair) internal damages automatically and autonomously. Definitely, it is a truly amazing characteristic that can fill not only the cracks but also small pinholes. The different types of materials including polymers, metals, and ceramics can have the ability to self-healing with their own self-healing mechanisms [19]. In this chapter, the focus is on self-healing polymers and their mechanisms. Different strategies can be considered to design self-healing polymers, such as the release of healing agents, reversible cross-links, and miscellaneous technologies, including electrohydrodynamics, shape memory effect, and nanoparticle migration. In this section, the different strategies are discussed.

4.1 Release of healing agents

The release of healing agents is one of the most covenant strategies for designing self-healing materials. This is accomplished by loading one or more healing agents, including monomers, dyes, catalysts, and hardeners in a container such as microcapsules, hollow fibers, and microvascular. The containers are then embedded into a polymeric matrix, which can release the healing agents by stimulation, such as physical damage or even pH alteration [20]. **Figure** 7 depicts a schematic of the healing process using the release of healing agents from containers.

4.1.1 Microencapsulation

In the encapsulation process, self-healing microcapsules are synthesized by loading a self-healing agent (solids, liquid, or gases) in an inert shell which in turn isolates and protects them from the external environments. Therefore, self-healing microcapsules are made of two parts, the core (healing agent) and the shell, which may vary in shape (spherical or irregular shape) and size (from nano to micro). **Figure 8** shows a schematic of a microcapsule structure. Generally, microcapsules have a diameter of 3–800 μm, and consist of 10–90 wt% core materials [23]. Different types of materials can be used as the healing agent and shell. Core materials or healing agents are selected based on the application of healing material such as stabilization against environmental degradation, improvement of longtime efficiency, maintenance of non-toxicity of degradation products, and easy handling through solidification of liquid core [24]. Some of the most common materials used in self-healing coatings are cerium nitrate,

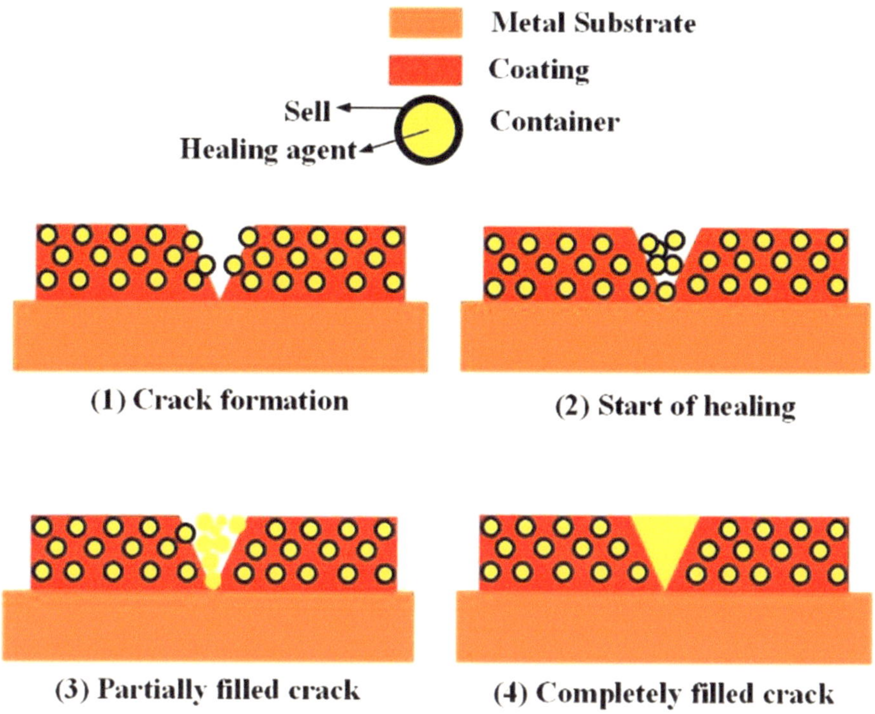

Figure 7.
A schematic of healing process using release of healing agents from containers [21].

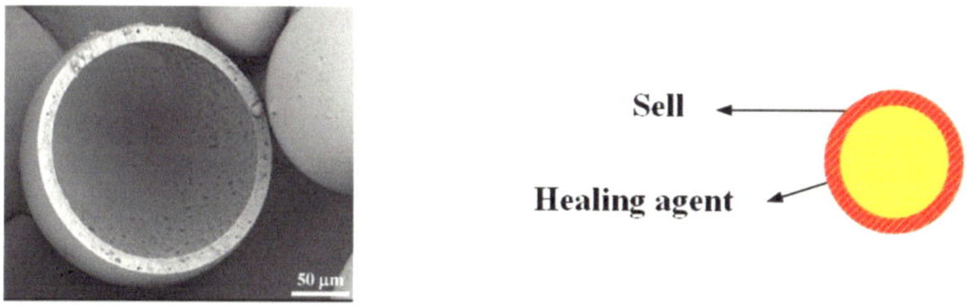

Figure 8.
A schematic of a microcapsule structure [22].

dodecylamine, polyethyleneimine, linseed oil, and sodium alginate. The shell material can be various from traditional organic polymers to novel inorganic materials. Usually, core materials are loaded in poly(Urea formaldehyde) (PUF), poly(melamine formaldehyde), cellulose nanofibers (CNF), halloysite nanotubes, etc. **Table 1** listed some conventional materials used in self-healing coating for different substrates.

Interfacial polymerization and in-situ polymerization are the most common methods for the synthesis of microcapsules, which are defined as follows [34]:

Interfacial polymerization: interfacial polymerization is an encapsulation procedure that mainly developed in the late 1960s [35]. This procedure consists of four main steps and two sets of monomers. One monomer is soluble in the oil phase and the other one is soluble in the water phase. At first monomer A (soluble in the water phase) is dissolved in the water phase. Secondly, monomer B (soluble in the oil phase) is dissolved in the oil phase. Then the oil phase is introduced into the water phase and emulsification is carried

Core Material	Metal substrate	containers	References
Cerium nitrate	Steel	PUF	[25]
Cerium	Aluminum alloy	Nanoporous silica	[26]
Sodium Alginate	Steel	Polyaniline	[27]
2-Mercaptobenzothiazole	Aluminum alloy	Silica nanocapsules	[28]
linseed oil	Mild steel	PUF	[29]
Oleic acid	Carbon steel	CNF	[30]
Silyl ester	Aluminum alloy	PUF	[31]
Dicyclopentadiene	—	PUF	[32]
Poly(dimethyl siloxane)	—	PUF	[33]

Table 1.
Some conventional materials used in self-healing coating for different substrates.

out under constant stirring. Finally, the polymerization reaction takes place through a chemical reaction between the monomers A and B which is initiated by changes in pH (acids or bases) and can be accelerated by the use of catalysts. The polymerization reaction leads to the form of a polymer film at the interface of monomers and the polymer formed is deposited around the drops which leads to encapsulation [35]. The encapsulation efficiency is enhanced when a low-molecular-weight shell material is used, due to the higher mobility of the small molecules. However, this reduces the shell's strength [36]. In addition, stirring speed is manipulated to control the size of the microcapsules. The higher the speed of the stirrer, the smaller the microcapsules are formed.

In-situ polymerization: encapsulation via in-situ polymerization technique is very similar to interfacial polymerization. The difference is that there are no reactants in the core material in *in situ* polymerization. In fact, in this process, polymerization takes place in the continuous phase, rather than in the interface between the continuous phase and the core material. In-situ polymerization also includes four steps. First, the core material is dispersed in the water phase. In the next step, the shell material is introduced into the water phase. Typically, this includes monomers that react in continuous phase and form a polymer. Then the pH is reduced by the addition of acid to initiate the polymerization reaction. Finally, the polymer film formed covers droplets. The size of the microcapsules is controlled by the agitator speed, so that the higher the speed of the stirrer, the smaller the microcapsules are achieved [37].

4.1.2 Hollow fiber embedment

The use of hollow fiber as a container for healing agents is another interesting method as agents can restore up to 97% of their initial flexural strength. In this method, the healing mechanism is similar to that of microcapsule-based methods with the difference that the healing agent is stored in hollow tubes or fibers until they are ruptured by damage [38]. **Figure 9** depicts a schematic of the self-healing concept using hollow fibers. Recently, fibers such as hollow glass fibers, hollow site nanotubes, titanium dioxide nanotubes, and polymeric fibers have been used as a container for self-healing applications [38]. In the hollow fibers approach, a healing agent and a curing agent (i.e., epoxy resin and its hardener) are loaded in separate hollow fibers that react together when both fibers are broken due to external damage.

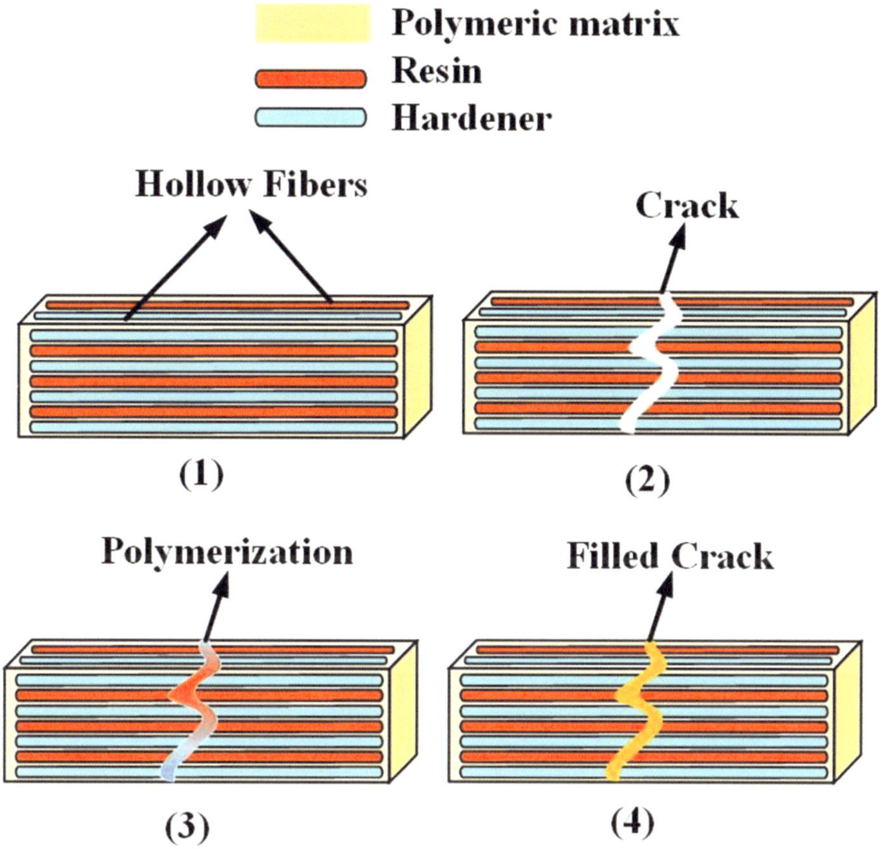

Figure 9.
A schematic of self-healing concept using hollow fibers [34].

One of the most remarkable advantages of this self-healing method is releasing a large amount of healing agent because of the large size of containers that can be suitable for filling cracks or large holes. However, this can also be a limitation. Because the rupture of hollow fibers depletes the healing agent contained within it which leads to a limitation on the number of times that a damaged region can be healed. On the other hand, this self-healing system faces a significant drawback which is the development of a practical technique for filling the hollow fibers with the liquid healing agent [39].

4.1.3 Microvascular system

To overcome the limitation of hollow fiber containers, microvascular networks have been introduced and developed by White et al. [40, 41] which are similar to the biological vascular system. In this method, the microvascular networks can repair the damaged area autonomously. Thanks to the small diameter of the microvascular, only the vascular in the damaged area will discharge the healing agent when damage occurs and the rest of the containers will remain intact. **Figure 10** shows a schematic of a microvascular network for self-healing applications. Although this method can increase the number of healing cycles, the fabrication process of such a microvascular

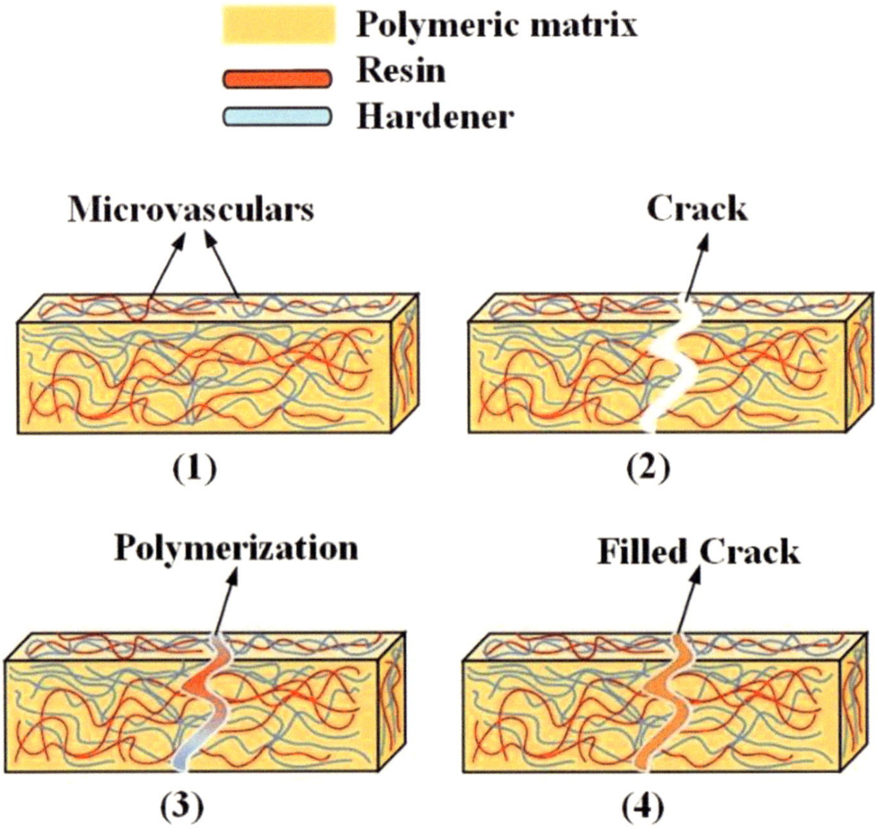

Figure 10.
A schematic of a microvascular network for self-healing application [41].

network is complex and it is hard to achieve synthetic materials with such properties for practical applications.

4.2 Reversible cross-links

In chemistry, cross-linking is the linking of one polymer chain to another one. The cross-linking is an irreversible process that can change the polymers' physical properties such as elasticity, mechanical behavior, and surface characteristics [42]. But highly cross-linked materials have the drawback of brittleness and tend to crack. Some approaches, including Diels–Alder (DA) and Retro-DA reactions as well as the use of ionomers and supramolecular polymers, are available to bring reversibility in cross-linked polymeric. Reversibly cross-linked polymers exhibit self-healing properties but they need an external trigger such as thermal or chemical activation. Thus, these systems show a non-autonomous healing phenomenon [34].

4.3 Miscellaneous technologies

Some other emerging technologies are available in the literature which are discussed in brief in this section. These methods have very limited applications and their explanation is beyond the goals of this chapter thus they are only mentioned.

These methods include electro hydrodynamics, conductivity, shape memory effect, nanoparticle migrations, and co-deposition [34].

5. Metal corrosion and self-healing coatings

As stated before, self-healing materials offer vast scientific attention due to the wide range of applications. In recent years, these new-generation protective coatings have been developed and commercialized for metal corrosion. In this regard, self-healable polymers find potential candidates for coatings due to their suitable thermo-mechanical behavior, lightweight, excellent adhesion, corrosion resistance, and good chemical resistance [43]. The major components of all coatings are the binders. So, it is obvious that the selection of the binder will affect the general performance of the coating. The most common binders are epoxy resin and polyurethane, however many other polymers, such as acrylic resins and polyesters, are used in many applications [44]. One of the most convenient methods proposed for self-healing coatings is adding containers loaded with a self-healing agent to a polymeric matrix (binders). Self-healing coatings can remarkably enhance the anti-corrosion performance and service life of metals. These protective coatings can restore the physical coating barrier, seal or close defects, or even inhibit the corrosion reactions at the coating defects. This is mostly accomplished by adding microcapsules loaded by a polymerizable healing agent or corrosion inhibitors to routine binders. The other self-healing design strategies are less commonly reported and therefore they are not discussed in this chapter. It is also worth noting that the coatings described here are mainly epoxy- or polyurethane-based organic systems. Two main classifications can consider for self-healing coating, autonomous and non-autonomous. In the following section, these two mechanisms are discussed in detail.

5.1 Autonomous healing mechanisms

Autonomous self-healing coating can repair their bulk integrity or functional properties without any external physical intervention. The Autonomous healing ability is usually achieved by embedding extrinsic polymerizable healing agents in the coating. In this system healing agents are mostly stored in microcapsules and released when the coating is damaged, which then polymerizes to fill the damage or form a protective film that inhibits the electrochemical reactions occurring at the exposed metal substrate [18].

5.1.1 Self-healing based on defect-filling effect

In this type of healing mechanism, the stored polymerizable healing agent releases due to generating a defect caused by mechanical damage. These materials can polymerize into a film by reacting with hardener or even with moisture or oxygen in the environment and can fill the defect. One of the classic examples of this type of healing mechanism was reported by White and Sottos [32] in 2001. In their proposed system, healing was accomplished by incorporating microencapsulated dicyclopentadiene in the PUF shell within an epoxy matrix. The average healing efficiency was reported 60% of the original fracture load. This method was further developed in the following years to be used for various metal substrates. **Table 2** listed several autonomous self-healing coatings based on defect-filling effects for different applications.

Type of container	Shell material	Core material	Polymer matrix	Substrate	References
Microcapsule	tris (p-isocyanatophenyl) thiophosphate	Isophorone diisocyanate	Epoxy	Carbon steel	[45]
Microcapsule	Urea-formaldehyde	Tung oil	Epoxy	Carbon steel	[46]
Microcapsule	Urea-formaldehyde	Maleic anhydride	Epoxy	Aluminum (A2024)	[47]
Hollow fiber	Halloysite nanotubes	Ce^{3+}/Zn^{4+}	Silica	Aluminum (A356)	[48]
Microcapsule	Polyurea	linseed oil	Polyurethane	Steel	[49]
Microcapsule	PUF	linseed oil	Epoxy	Magnesium	[50]
Hollow fiber	TiO_2 nanotubes	Amine	Epoxy	Carbon steel	[51]
Microcapsule	PUF	Cerium Nitrate	Epoxy	Carbon steel	[52]

Table 2.
Several autonomous self-healing coatings based on defect-filling effects for different applications.

5.1.2 Self-healing based on corrosion inhibitors

The embedding of corrosion inhibitors in coatings is another mechanism of autonomous self-healing systems. Some of the most widely used corrosion inhibitors in autonomous self-healing coating are nitrites, phosphates, vanadates, molybdates, tungstates, borates, mercaptobenzothiazole, benzotriazole, imidazoline, 8-hydroxyquinoline, and aliphatic amines [18]. In this type of self-healing coatings, the healing process is accomplished by anodic dissolution and cathodic reactions which leads to corrosion inhibition. The conceptual design of inhibitor-based coatings is simple and this can be achieved by adding corrosion inhibitors directly to a polymeric matrix. In the early studies, the addition of doping nitrates and phosphates of cerium into organic coatings was proposed to prevent the corrosion of zinc, galvanized steel, and aluminum alloy [53–56]. This method is subject to problems such as poor compatibility between the organic coating resins and particle agglomeration. However, recently, the approach of encapsulating inhibitor agents has received more attention because in this way inhibitors can be released in a stable and controlled manner [57]. **Table 3** listed a number of self-healing based on corrosion inhibitors for different applications.

5.2 Non-autonomous healing mechanisms

In non-autonomous systems, healing effects accomplish by an external stimuli, such as heat and light, which trigger the chemical reactions or physical transitions necessary for bond formation or molecular chain movement. In fact, this type of coating are healed by recovering the intrinsic chemical bonds and/or physical configurations of the polymer networks in the coating matrices. The external stimulus provides the activation energy required for bond breakage/reformation. For example a heat stimulus can enhance the reactions of the broken bonds by bringing them

Type of container	Shell material	Core material	Polymer matrix	Substrate	References
Nanocontainer	Ceramic	2-mercaptobenzothiazole	Epoxy	Galvanized steel	[58]
Nanocontainer	Cerium molybdate	8-hydroxyquinoline	Polystyrene	Aluminum (AA2024)	[59]
Directly added	—	hydrotalcite	Polyvinyl butyral	Aluminum (AA2024)	[60]
Embedded container	Porous graphene	Benzotriazole	Epoxy	Carbon steel	[61]
Nanocontaine	(2D) covalent organic framework	Benzotriazole	Epoxy	Carbon steel	[62]
Directly added	Microfiber	Superabsorbent polymer	Vinyl-ester	Carbon steel	[63]

Table 3.
A number of self-healing based on corrosion inhibitors for different application.

closer together. The most common light stimulus are sunlight, near infrared (NIR) light, and UV light. But heat sources can be artificially applied (e.g., by a heat gun) or generated from the service environments (e.g., sunlight, abrasion) [18]. Non-autonomous healing polymers have vast applications in healthcare, aerospace, construction and electronics industries. Thus, a comprehensive review of these materials beyond the scope of this chapter and only some systems that are more closely related to protective coatings are discussed in the following section.

5.2.1 Non-autonomous self-healing based on dynamic bonds

The dynamic bonds refer to reversible break and reform bonds which allow a continuous modification of the constitution by reorganization and exchange of building blocks [64]. For instance, at a certain temperature thermally reversible bonds can decompose, which allows the polymer chains to flow to the defect and re-crosslink to repair the defect [65]. Diels-Alder (DA) reaction is one of the most prominent examples of this healing system. A recent work by Chuo et al. [66] demonstrated the possibility of a tetra-functional furan-capped aniline trimer, a trifunctional maleimide, and a trifunctional to prevent metal corrosion. The corrosion current density in the polarization curve was used to evaluate the corrosion protection efficiency of the scratched coating. The results of a cycle test showed that the proposed system could recover protection efficiency from 79.8% to 99.2%. Some other researchers reported self-healing effects by using light stimuli. For example, UV-sensitive self-healing polymers have been developed based on reversible photo-crosslinking reactions [67], or near-infrared light-triggered self-healing ability was used for biocomposites [68]. Light-responsive self-healing polymers have several major advantages over thermally induced self-healing systems, e.g., the healing can be triggered instantaneously, remotely, and on demand. In fact, unlike heat-stimuli systems where all surfaces are exposed, in light-responsive systems, the light stimulus can be applied exactly to the damaged area, which leads to a reduction in side reactions and degradation in the intact coating during the healing process [18].

5.2.2 Non-autonomous self-healing based on shape memory polymers (SMPs)

Shape memory materials are a kind of smart materials that can recover their original shape from a deformed state by applying external stimuli such as heat or light. Both polymers and alloys can perform shape memory behaviors with different mechanisms. In shape memory alloys (SMAs), such as NiTi-based, Cu-based (CuAlNi and CuZnAl), and Fe-based alloys, the shape memory effects are governed by the phase transformation among twinned martensite, detwinned martensite, and austenite [69]. In shape memory polymers (SMPs), the shape memory effects are usually due to the viscoelastic transformation of polymer chains when cycled through a thermal transition temperature, such as a glass transition temperature (Tg) or a melting temperature (Tm) [70]. From the engineering point of view, tailoring the properties is much easier for polymers than metals/alloys. Polymers have traditionally lower material prices and processing costs [71]. Furthermore, SMPs have lower weight, easy fabrication methods, and higher elasticity. Hence, they have been proposed for biomedical [72], aerospace [73], and many other applications.

6. Conclusion

This chapter provides brief information on metal corrosion and corrosion prevention methods. Some of the most common metal corrosion mechanisms, such as uniform, galvanic, crevice, pitting, and erosion-corrosion, as well as mechanically assisted corrosion, are introduced. In addition, various methods of preventing metal corrosion, such as cathodic protection, galvanizing, painting, and coatings, are discussed and their advantages and disadvantages are addressed. A summary of the information mentioned in this chapter is summarized here.

Investigations show that corrosion is an inevitable and costly phenomenon, and its prevention is of great importance. The Cathodic protection as a classic method is efficient, but it is expensive and requires skilled workers and manpower as well as periodic inspections. Galvanizing is also an effective method but has only limited applications for steel. Furthermore, the galvanized coating can easily chip off and create an uneven surface that leads to corrosion and possible loss of life. Painting is also mainly decorative and does not help to prevent corrosion effectively. Paints can easily deteriorate due to environmental factors or mechanical damage and expose the surface to corrosive factors. It can be said that coatings are the most efficient method to prevent corrosion as they are cost-effective and provide a significant barrier to corrosive factors. Definitely, coatings prolong the service life of metals, but they face the problem of cracking defects and mechanical damage. However, a new generation of coatings called self-healing coatings has been introduced and developed in recent years, which has solved the mentioned problems.

Self-healing materials can heal (recover/repair) internal damages automatically and autonomously. Different design strategies such as the release of healing agents, reversible cross-links, and miscellaneous technologies including electrohydrodynamics, and shape memory effect are available for self-healing fabrication. Release of healing agents is probably the most convenient method to fabricate self-healing coatings in which healing agents are loaded in a container and added to a polymer matrix, and when damage occurs, it releases and repairs the damaged area. This method is widely used and effective because the healing agent can be released in a stable and controlled manner. Two main classifications can consider for self-healing coating, autonomous and non-autonomous. Autonomous self-healing coating can repair their bulk integrity or functional properties without any external physical intervention. But In non-autonomous systems, healing effects accomplish by external stimuli such as heat and light. Both mechanisms are very important and have their own applications despite their few limitations.

Considering the above mentioned properties, self-healing coating are a promising method for prevention of metal corrosion and they have revolutionized the coating systems. Although they are still in the early stages of their development, they have shown a great potential for wide applications and seems to completely replace traditional coatings in the future.

Conflict of interest

The authors declare no conflict of interest.

Nomenclature

2D	Two dimensional
CNF	Cellulose nanofibers
DA	Diels–Alder
DC	Direct current
ICCP	Impressed current cathodic protection
NIR	Near infrared
PUF	Poly(Urea formaldehyde(
SACP	sacrificial anodes cathodic protection
SMAs	shape memory alloys
SMPs	shape memory polymers
UV	Ultraviolet

Author details

Shahin Kharaji
FAPKCO Industrial Company, Shiraz, Iran

*Address all correspondence to: shahinkharaji@gmail.com

References

[1] Palanisamy G. Corrosion inhibitors. In: Singh A, editor. Corrosion Inhibitors [Internet]. London: IntechOpen; 2019 [cited 2022 Dec 29]. Available from: https://www.intechopen.com/chapters/64392. DOI: 10.5772/intechopen.80542

[2] Foorginezhad S, Mohseni-Dargah M, Firoozirad K, Aryai V, Razmjou A, Abbassi R, et al. Recent advances in sensing and assessment of corrosion in sewage pipelines. Process Safety and Environmental Protection. 2021;**147**:192-213

[3] Shreir LL. Corrosion: Metal/Environment Reactions. Oxford, UK: Newnes; 2013

[4] Gardner L. Stability and design of stainless steel structures–review and outlook. Thin-Walled Structures. 2019;**141**:208-216

[5] Dehghani A, Aslani F, editors. A Review on Defects in Steel Offshore Structures and Developed Strengthening Techniques. Structures. London, United Kingdom: The Institution of Structural Engineers International HQ, Elsevier; 2019

[6] Chaves I, Melchers R. Coastal and offshore structures, wave and wind loading microstructural effect on the marine corrosion of low-carbon steel weldments. In: Mechanics of Structures and Materials XXIV. London, England: CRC Press; 2019. pp. 898-903

[7] Senior NA, Martino T, Diomidis N, Gaggiano R, Binns J, Keech P. The measurement of ultra low uniform corrosion rates. Corrosion Science. 2020;**176**:108913

[8] Paglia C, Antonietti S, Mosca C, editors. The deterioration of 100 years old coated steel bridges. In: IOP Conference Series: Materials Science and Engineering, Volume 1150, International Conference on Material Science and Engineering Technology (ICMSET 2021), 12th-14th March 2021, Jeju Island, Korea. IOP Publishing; 2021

[9] Akhoondan M, Bell GE. Fastener corrosion. Structure. 2016;**74**:74

[10] Zadorozne NS et al. Crevice corrosion kinetics of nickel alloys bearing chromium and molybdenum. Electrochimica Acta. 2012;**76**:94-101

[11] During ED. Corrosion Atlas: A Collection of Illustrated Case Histories. Regentesselaan, Amersfoort, The Netherlands: Association of Companies and Organizations for Energy and Environment, Elsevier; 2018

[12] Medvedovski E. Formation of corrosion-resistant thermal diffusion boride coatings. Advanced Engineering Materials. 2016;**18**(1):11-33

[13] Brownlie F, Giourntas L, Hodgkiess T, Palmeira I, Odutayo O, Galloway A, et al. Effect of cathodic protection methods on ferrous engineering materials under corrosive wear conditions. Corrosion Engineering, Science and Technology. 2020;**55**(6):480-486

[14] Tezdogan T, Demirel YK. An overview of marine corrosion protection with a focus on cathodic protection and coatings. Brodogradnja: Teorija i praksa brodogradnje i pomorske tehnike. 2014;**65**(2):49-59

[15] Shibli S, Meena B, Remya R. A review on recent approaches in the field of hot dip zinc galvanizing process. Surface and Coatings Technology. 2015;**262**:210-215

[16] Kendig M, Mills DJ. An historical perspective on the corrosion protection by paints. Progress in Organic Coatings. 2017;**102**:53-59

[17] Popoola A, Olorunniwo O, Ige O. Corrosion resistance through the application of anti-corrosion coatings. Developments in Corrosion Protection. 2014;**13**(4):241-270

[18] Zhang F, Ju P, Pan M, Zhang D, Huang Y, Li G, et al. Self-healing mechanisms in smart protective coatings: A review. Corrosion Science. 2018;**144**:74-88

[19] Wool RP. Self-healing materials: A review. Soft Matter. 2008;**4**(3): 400-418

[20] Yuan Y, Yin T, Rong M, Zhang M. Self healing in polymers and polymer composites. Concepts, realization and outlook: A review. Express Polymer Letters. 2008;**2**(4):238-250

[21] Huang Y et al. Triple-action self-healing protective coatings based on shape memory polymers containing dual-function microspheres. ACS Applied Materials & Interfaces. 2018;**10**(27):23369-23379

[22] You X, Wang B, Xie S, Li L, Lu H, Jin M, et al. Microfluidic-assisted fabrication of monodisperse Core–Shell microcapsules for pressure-sensitive adhesive with enhanced performance. Nanomaterials. 2020;**10**(2):274

[23] Samadzadeh M, Boura SH, Peikari M, Kasiriha S, Ashrafi A. A review on self-healing coatings based on micro/nanocapsules. Progress in Organic Coatings. 2010;**68**(3):159-164

[24] Chao D. The role of surfactants in synthesizing polyurea microcapsule. Journal of applied Polymer Science. 1993;**47**(4):645-651

[25] Matsuda T, Jadhav N, Kashi KB, Jensen M, Suryawanshi A, Gelling VJ. Self-healing ability and particle size effect of encapsulated cerium nitrate into pH sensitive microcapsules. Progress in Organic Coatings. 2016;**90**:425-430

[26] Jiang X, Jiang Y-B, Liu N, Xu H, Rathod S, Shah P, et al. Controlled Release from Core-Shell Nanoporous silica particles for corrosion inhibition of aluminum alloys. Journal of Nanomaterials. 2010;**2011**:15

[27] Cui J, Li X, Pei Z, Pei Y. A long-term stable and environmental friendly self-healing coating with polyaniline/sodium alginate microcapsule structure for corrosion protection of water-delivery pipelines. Chemical Engineering Journal. 2019;**358**:379-388

[28] Montemor M, Snihirova D, Taryba M, Lamaka S, Kartsonakis I, Balaskas A, et al. Evaluation of self-healing ability in protective coatings modified with combinations of layered double hydroxides and cerium molibdate nanocontainers filled with corrosion inhibitors. Electrochimica Acta. 2012;**60**:31-40

[29] Suryanarayana C, Rao KC, Kumar D. Preparation and characterization of microcapsules containing linseed oil and its use in self-healing coatings. Progress in Organic Coatings. 2008;**63**(1):72-78

[30] Yabuki A, Shiraiwa T, Fathona IW. pH-controlled self-healing polymer coatings with cellulose nanofibers providing an effective release of corrosion inhibitor. Corrosion Science. 2016;**103**:117-123

[31] García S, Fischer H, White P, Mardel J, González-García Y, Mol J, et al. Self-healing anticorrosive organic coating based on an encapsulated water reactive silyl ester: Synthesis and proof

of concept. Progress in Organic Coatings. 2011;**70**(2-3):142-149

[32] White SR, Sottos NR, Geubelle PH, Moore JS, Kessler MR, Sriram S, et al. Autonomic healing of polymer composites. Nature. 2001;**409**(6822):794-797

[33] Keller MW, White SR, Sottos NR. An elastomeric self-healing material. In: Proceedings of the 2006 SEM Annual Conference and Exposition on Experimental and Applied Mechanics, Saint Louis, MO, USA, 4-7 June 2006. Vol. 1. pp. 379-382

[34] Ghosh SK. Self-Healing Materials: Fundamentals, Design Strategies, and Applications. Minneapolis, United States: Wiley Online Book, Corporate Headquarters, Wiley Online Library; 2009

[35] Perignon C, Ongmayeb G, Neufeld R, Frere Y, Poncelet D. Microencapsulation by interfacial polymerisation: Membrane formation and structure. Journal of Microencapsulation. 2015;**32**(1):1-15

[36] Al-Shannaq R, Farid MM. Microencapsulation of phase change materials for thermal energy storage systems. In: Advances in Thermal Energy Storage Systems. Cambridge, United Kingdom: Woodhead Publishing Limited, Elsevier; 2021. pp. 269-329

[37] Nelson G. Microencapsulated colourants for technical textile application. In: Advances in the Dyeing and Finishing of Technical Textiles. Cambridge, United Kingdom: Woodhead Publishing Limited, Elsevier; 2013. pp. 78-104

[38] Pulikkalparambil H et al. Self-repairing hollow-fiber polymer composites. Self-Healing Composite Materials. 2020;**1**:313-326

[39] Banea MD, da Silva LF, Campilho RD, Sato C. Smart adhesive joints: An overview of recent developments. The Journal of Adhesion. 2014;**90**(1):16-40

[40] Therriault D, White SR, Lewis JA. Chaotic mixing in three-dimensional microvascular networks fabricated by direct-write assembly. Nature Materials. 2003;**2**(4):265-271

[41] Toohey KS, Sottos NR, Lewis JA, Moore JS, White SR. Self-healing materials with microvascular networks. Nature Materials. 2007;**6**(8):581-585

[42] Wang K, Amin K, An Z, Cai Z, Chen H, Chen H, et al. Advanced functional polymer materials. Materials Chemistry Frontiers. 2020;**4**(7):1803-1915

[43] Cho SH, White SR, Braun PV. Self-healing polymer coatings. Advanced Materials. 2009;**21**(6):645-649

[44] Pulikkalparambil H, Siengchin S, Parameswaranpillai J. Corrosion protective self-healing epoxy resin coatings based on inhibitor and polymeric healing agents encapsulated in organic and inorganic micro and nanocontainers. Nano-Structures & Nano-Objects. 2018;**16**:381-395

[45] Attaei M, Calado LM, Taryba MG, Morozov Y, Shakoor RA, Kahraman R, et al. Autonomous self-healing in epoxy coatings provided by high efficiency isophorone diisocyanate (IPDI) microcapsules for protection of carbon steel. Progress in Organic Coatings. 2020;**139**:105445

[46] Samadzadeh M, Boura SH, Peikari M, Ashrafi A, Kasiriha M. Tung oil: An autonomous repairing agent for self-healing epoxy coatings. Progress in Organic Coatings. 2011;**70**(4):383-387

[47] Njoku CN, Arukalam IO, Bai W, Li Y. Optimizing maleic anhydride microcapsules size for use in self-healing epoxy-based coatings for corrosion protection of aluminum alloy. Materials and Corrosion. 2018;**69**(9):1257-1267

[48] Manasa S, Jyothirmayi A, Siva T, Sathiyanarayanan S, Gobi K, Subasri R. Effect of inhibitor loading into nanocontainer additives of self-healing corrosion protection coatings on aluminum alloy A356. 0. Journal of Alloys and Compounds. 2017;**726**:969-977

[49] Tatiya PD, Hedaoo RK, Mahulikar PP, Gite VV. Novel polyurea microcapsules using dendritic functional monomer: Synthesis, characterization, and its use in self-healing and anticorrosive polyurethane coatings. Industrial & Engineering Chemistry Research. 2013;**52**(4):1562-1570

[50] Alias J, Johari N, Zanurin A, Alang N, Zain M, editors. Self-healing epoxy coating with microencapsulation of linseed oil for the corrosion protection of magnesium (Mg). In: Journal of Physics: Conference Series, Volume 2129, 1st International Conference on Material Processing and Technology (ICMProTech 2021), 14th-15th July 2021, Perlis, Malaysia. IOP Publishing; 2021

[51] Vijayan PP, Al-Maadeed MAS. TiO2 nanotubes and mesoporous silica as containers in self-healing epoxy coatings. Scientific Reports. 2016;**6**(1):1-9

[52] Farzi G, Davoodi A, Ahmadi A, Neisiany RE, Anwer MK, Aboudzadeh MA. Encapsulation of cerium nitrate within poly (urea-formaldehyde) microcapsules for the development of self-healing epoxy-based coating. ACS Omega. 2021;**6**(46):31147-31153

[53] Aramaki K. Self-healing mechanism of an organosiloxane polymer film containing sodium silicate and cerium (III) nitrate for corrosion of scratched zinc surface in 0.5 M NaCl. Corrosion Science. 2002;**44**(7):1621-1632

[54] Trabelsi W, Cecilio P, Ferreira M, Montemor M. Electrochemical assessment of the self-healing properties of Ce-doped silane solutions for the pre-treatment of galvanised steel substrates. Progress in Organic Coatings. 2005;**54**(4):276-284

[55] Mardel J, Garcia S, Corrigan P, Markley T, Hughes A, Muster T, et al. The characterisation and performance of Ce (dbp) 3-inhibited epoxy coatings. Progress in Organic Coatings. 2011;**70**(2-3):91-101

[56] Montemor M, Pinto R, Ferreira M. Chemical composition and corrosion protection of silane films modified with CeO2 nanoparticles. Electrochimica Acta. 2009;**54**(22):5179-5189

[57] Choi H, Kim KY, Park JM. Encapsulation of aliphatic amines into nanoparticles for self-healing corrosion protection of steel sheets. Progress in Organic Coatings. 2013;**76**(10):1316-1324

[58] Kartsonakis I, Balaskas A, Koumoulos E, Charitidis C, Kordas G. Incorporation of ceramic nanocontainers into epoxy coatings for the corrosion protection of hot dip galvanized steel. Corrosion Science. 2012;**57**:30-41

[59] Kartsonakis IA, Kordas G. Synthesis and characterization of cerium molybdate nanocontainers and their inhibitor complexes. Journal of the American Ceramic Society. 2010;**93**(1):65-73

[60] Williams G, McMurray HN. Inhibition of filiform corrosion on polymer coated AA2024-T3 by

hydrotalcite-like pigments incorporating organic anions. Electrochemical and Solid-State Letters. 2004;7(5):B13

[61] Ye Y, Chen H, Zou Y, Ye Y, Zhao H. Corrosion protective mechanism of smart graphene-based self-healing coating on carbon steel. Corrosion Science. 2020;**174**:108825

[62] Liu T, Li W, Zhang C, Wang W, Dou W, Chen S. Preparation of highly efficient self-healing anticorrosion epoxy coating by integration of benzotriazole corrosion inhibitor loaded 2D-COF. Journal of Industrial and Engineering Chemistry. 2021;**97**:560-573

[63] Yabuki A, Tanabe S, Fathona IW. Self-healing polymer coating with the microfibers of superabsorbent polymers provides corrosion inhibition in carbon steel. Surface and Coatings Technology. 2018;**341**:71-77

[64] Chakma P, Konkolewicz D. Dynamic covalent bonds in polymeric materials. Angewandte Chemie. 2019;**131**(29):9784-9797

[65] García S, Fischer H, Van Der Zwaag S. A critical appraisal of the potential of self healing polymeric coatings. Progress in Organic Coatings. 2011;**72**(3):211-221

[66] Chuo T-W, Liu Y-L. Furan-functionalized aniline trimer based self-healing polymers exhibiting high efficiency of anticorrosion. Polymer. 2017;**125**:227-233

[67] Habault D, Zhang H, Zhao Y. Light-triggered self-healing and shape-memory polymers. Chemical Society Reviews. 2013;**42**(17):7244-7256

[68] Xiong S, Wang Y, Zhu J, Yu J, Hu Z. Poly (ε-caprolactone)-grafted polydopamine particles for biocomposites with near-infrared light triggered self-healing ability. Polymer. 2016;**84**:328-335

[69] Zareie S, Issa AS, Seethaler RJ, Zabihollah A, editors. Recent Advances in the Applications of Shape Memory Alloys in Civil Infrastructures: A Review. Structures. London, United Kingdom: The Institution of Structural Engineers International HQ, Elsevier; 2020

[70] Hornat CC, Urban MW. Shape memory effects in self-healing polymers. Progress in Polymer Science. 2020;**102**:101208

[71] Huang W, Ding Z, Wang C, Wei J, Zhao Y, Purnawali H. Shape memory materials. Materials Today. 2010;**13**(7-8):54-61

[72] Chan BQY, Low ZWK, Heng SJW, Chan SY, Owh C, Loh XJ. Recent advances in shape memory soft materials for biomedical applications. ACS Applied Materials & Interfaces. 2016;**8**(16):10070-10087

[73] Sun J, Guan Q, Liu Y, Leng J. Morphing aircraft based on smart materials and structures: A state-of-the-art review. Journal of Intelligent Material Systems and Structures. 2016;**27**(17):2289-2312

Section 3

Corrosion and Protection

Chapter 6

Corrosion Fatigue Behavior and Damage Mechanism of the Bridge Cable Structures

Guowen Yao, Xuanbo He, Jiawei Liu, Jiangshan Lu and Zengwei Guo

Abstract

The long-term performance and corrosion fatigue damage status were investigated and analyzed under the service environment for the cable structures in cable-stayed bridges, suspension bridges, and suspender arch bridges. The artificial accelerated corrosion fatigue tests were carried out on galvanized parallel steel wire under coupled loading and environments. The damage mechanisms of galvanized parallel steel wire in corrosion, stress corrosion, and corrosion fatigue were investigated. The change laws of the mechanical properties of the cable were studied. Based on the image gray analysis, the evaluation method was proposed for the technical status of the damaged cable. Furthermore, combined with the cable damage evolution model, the service life prediction method and assessment technology of cables based on damage safety are established.

Keywords: bridge engineering, corrosion fatigue, cable structure, steel wire, damage

1. Introduction

The construction of large-span cable-stayed bridges around the world is at its peak, and a large number of cable-stayed bridges are already in the operational maintenance phase. The cable-stayed structure is one of the main load-bearing structures of cable-stayed bridges, and its durability directly affects the safety of bridge operation. As the cable system is subjected to long-term alternating loads and exposed to the natural environment, it is highly susceptible to environmental erosion. In particular, the acid rain area in China has accounted for one-third of the national territory, becoming the third largest heavy acid rain deposition area in the world after Europe and North America. In such areas with serious atmospheric pollution and water pollution or marine environment, the steel wire inside the cable-stayed bridge is highly susceptible to erosion by corrosive media, such as SO_4^{2-} and Cl^- in the environment. When the wire inside the cable breaks due to severe corrosion, it can lead to the failure of the cable system, affecting the safety and service life of the bridge, and the dangerous situation can cause serious economic losses and safety accidents. Therefore, cable corrosion and wire breakage have become the problems faced by most cable-stayed bridges.

In this chapter, the investigation and statistical analysis of the working performance and corrosion fatigue damage of cable under service environment are carried out for cable-stayed bridges, suspension bridges, arch bridges, and other bridge cable structures, and indoor artificially accelerated corrosion fatigue test research under the coupling effect of load and environment is conducted to reveal the corrosion fatigue damage mechanism of galvanized parallel wire cable under the coupling effect of salt spray environment and static stress and alternating stress. The physical and mechanical properties of the damaged cables were studied. The technical condition assessment method based on the image grayscale analysis is established and further combined with the cable damage evolution model, the service life prediction method and assessment technology based on the breakage safety of cable are established to provide a theoretical basis for the scientific evaluation of the technical condition of cable-stayed bridges, prolonging the service life of the cable and improving the anti-corrosion design of the cable.

2. Investigation and analysis of corrosion and fatigue conditions of bridge cable structures under service environment

The survey of cable-stayed bridges built around the world shows that cable-stayed bridges are facing serious challenges and threats from cable durability issues (**Figure 1**). For example, the South Pan River Bridge in Guizhou, the original main bridge was a 240-meter span steel truss suspension bridge, which was put into operation in November 1998 and has been in operation for less than 17 years, the main cable and sling suffered from severe corrosion was dismantled in 2015 and rebuilt as a continuous rigid structure bridge. Built in 2012, the Ximo River Bridge on the county highway from Zamu Town to Murdo in Bomi County, Tibet was also facing replacement when its main cable and sling were found to be seriously corroded during an inspection in 2017. The cable of cable-stayed bridges and the suspenders of half-through and through-arch bridges are more frequently replaced due to corrosion damage and reduced reliability. The St. Nazaire Bridge in France suffered from severe corrosion during its service life, with extensive rusting and spalling of the cable

Figure 1.
Corrosion and fatigue damage of cable-stayed bridges in mountainous areas with cable-stayed cable, suspension cable, and main cable.

surfaces. After two years of operation, the Kohlbrand Estuary Bridge in Germany suffered from severe corrosion at the lower anchorage end of its cable, with 25 broken wires in the cable, and the entire bridge was replaced at the cost of $60 million or four times as much as original cost. The cost of replacing the corroded cable-stay on the Maracaibo Bridge in Venezuela is also about $50 million.

Japan and the United States have conducted experimental and theoretical studies on the corrosion of cable for a long time, especially the corrosion of cable in the marine environment. The Honshu-Shikoku Bridge Management Department in Japan has accumulated years of survey data through on-site investigations and detailed knowledge of the damage caused by corrosion of cable in several cable-stayed and suspension bridges during operation and has conducted research on preventive maintenance methods for cable, proposing measures such as anticorrosion coating coverage, ventilation, and dehumidification, as well as static and dynamic behavior monitoring [1]. Kheyroddin [2] used seawater as a corrosive medium and carried out accelerated corrosion tests under static stress to comparatively study the anticorrosion effect of different protection systems. In addition, Barton [3] focused on the corrosion behavior of lasso in NaCl corrosive medium by artificially accelerated corrosion experiment, obtaining the weight loss of steel wire samples, hydrogen concentration, elongation, etc. The results of the study showed that water and temperature are the environmental factors most likely to cause corrosion of lasso, and accelerated corrosion test proves that the most serious corrosion occurs at the anchored end of lasso, the middle and the top end is less affected by corrosion, and the increase of NaCl concentration in the solution and temperature will accelerate the corrosion process of galvanized steel strands. In 1987, Dolley [4] and Mahmoud [5] surveyed nearly 100 cable-stayed bridges around the world and conducted a visual inspection of the bridges, finding that "nearly 200 cable-stayed bridges built worldwide in the last few decades are at risk due to cable corrosion." The new parallel wire system that was introduced in Japan also has a corrosion resistance life of only 25–30 years. Cable-stayed bridges are experiencing serious challenges and threats from cable durability issues [6–10].

3. Artificially accelerated corrosion fatigue experiment under coupled loading and environment

3.1 Experiment content and its implementation

Accelerated corrosion experiment was conducted on lassos using an acidic salt spray environment. The cable consisted of high-strength galvanized parallel steel wire. The strength level of the lasso is 1860 MPa, the yield strength is 1660 MPa, and the diameter φ is 5.2 mm. The loading force for both alternating stress and static stress is 1100 MPa, which is close to the stress corrosion threshold. At the end of the experiment, the steel wire was wiped with 10% dilute sulfuric acid to remove the surface corrosion, the mass loss of the steel wire subjected to corrosion was measured, and the mechanical properties of the corroded steel wire were tested to discuss the mechanical property response of the steel wire to the corrosive environment and to provide an experimental data basis for evaluating the corrosion damage and evolution of the steel wire. The main experimental equipment is shown in **Table 1**.

The salt solution was 5% NaCl solution, adding concentrated sulfuric acid in the solution to adjust the pH to 1. The experiment temperature reference document [11]

Name	Model	Technical parameters	Quantity (units)
Salt spray test chamber	YC-200	Salt spray deposition: 250 ml/m^2/h	1
Electro-hydraulic servo universal testing machine	WAW-1000	Maximum load: 1000 kN	1
Electronic balance	SL500ZN	Precision: 0.01 g	1
Double-acting hollow jack	ZKD	Carrying capacity: 20 t	4
Ultra-high-pressure electric oil pump	DSS	Power: 0.75 kW	1

Table 1.
Main experimental equipment.

will be set at 50°C ± 2°C, the air pressure is controlled between 70–170 kPa. Spray volume reference document [11] CASS experimental standards, taken as 250 ml/m^2/h. The specific implementation steps of experiment are as follows: All the treated steel wire were weighed with a balance, the diameters were measured, and the raw data were recorded. The experiment was arranged in three loading methods: stress-free, static stress, and alternating stress. The stress-free loading method cut the steel wire into 1.2 m sections, marked them, and put them directly into the corrosion chamber for continuous corrosion without interruption. When subjected to static stress loading, the steel wire is cut into 5.4 m sections, marked and through the corrosion box and the counter-weight table reserved hole, and anchored with a tensioning jack, the rest of the process is same as stress-free loading. Alternating stress loading, the same steel wire is cut into 5.4 m sections, the loading cycle for 4 h, that is, 2 h loading, 2 h unloading, the rest of the work is same as stress-free loading, every 24 h to open the corrosion box, take pictures, and record data, reconfigure and add solution. After the experiment, first, clean the corrosion products on the specimen with dilute sulfuric acid, then wash the steel wire with a large amount of water, blow dry with a blowing air and then weigh again, the diameter should be randomly measured three times for each specimen section, and take its average value and record. The last is the corrosion of steel wire mechanical tensile experiment.

In this experiment, at least five steel wires were used in each batch, two of which were alternately stressed, two were statically stressed, and the others were unstressed wires. The peak stress applied is 1100 MPa. The specific arrangement is shown in **Figure 2** [12].

Before conducting the formal tensile experiment, three intact steel wire samples with a length of 1 m were taken and tensile experiments were performed on them with a universal testing machine. Their elastic modulus E, tensile strength R_m, yield strength $R_{p0.2}$, and post-break elongation A were measured to determine the original mechanical properties parameters of the steel wire. It provides a reference for the test results of mechanical properties of steel wire after corrosion. The data obtained from the tensile experiment are given in **Table 2**.

In order to ensure the safety of the bridges, considering the safety factor of the steel wire used in cable-stayed bridges, the stresses applied during their operation should be less than their yield strength and tensile strength. Therefore, it is reasonable to apply a maximum tensile stress of 1100 MPa to the galvanized steel wire to ensure that they do not yield and fracture during the experimental process.

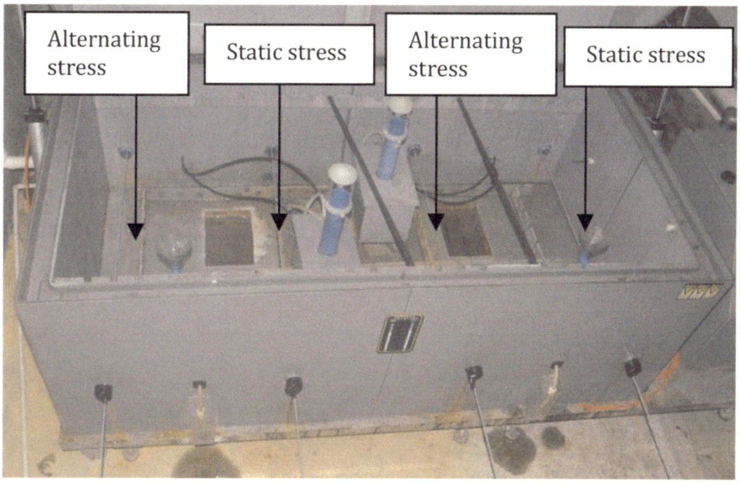

Figure 2.
Experimental specimen arrangement.

Number	E (GPa)	R_m (MPa)	$R_{p0.2}$ (MPa)	A (%)
1	205.54	1915	1760	5.5
2	231.36	1910	1710	6.0
3	213.20	1910	1730	4.5

Table 2.
Steel wire static performance experiment data.

3.2 Steel wire corrosion fatigue damage phenomenon

After the test of each batch of steel wire specimens placed in the corrosion test chamber, take them out to observe the morphology of corrosion products, and then remove the corrosion products. Wait for the drying of the steel wire and then observe its surface morphology, and preliminarily analyze the experimental phenomenon (**Figure 3**) [13], the conclusions are as follows:

1. Under the erosion of the corrosion solution, the corrosive medium in the solution gradually accumulates on the surface of the galvanized layer of the steel wire, and the galvanized layer undergoes nonuniform corrosion and gradually produces corrosion pits. The acidic corrosion solution penetrates into the corrosion pits, which continue to develop in depth and appear white corrosion, with the main corrosion product being $Zn(OH)_2$-$3ZnSO_4$–$5H_2O$. Due to stress, cracks appear in the galvanized layer of the steel wire surface, corrosion pits continue to expand and increase, the surface damage is further aggravated, the corrosive medium is more likely to penetrate into the galvanized layer, and the degree of corrosion of the steel wire continues to increase, which led to a large diameter of the corrosion pit on the surface of the steel wire. (**Figure 3a**);

2. With the further development of corrosion, white corrosion increases, and red corrosion is seen sporadically in the galvanized layer corrosion products. The area

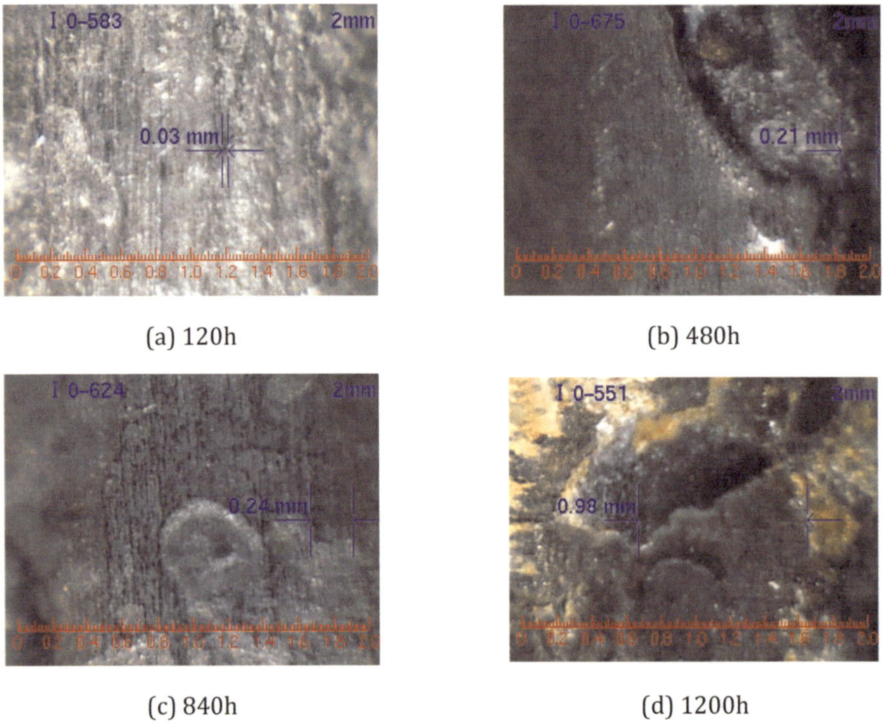

Figure 3.
Apparent condition of steel wire with different corrosion times [13] (a) 120 h (b) 480 h (c) 840 h (d) 1200 h.

and depth of the corrosion pits increase. After removing the corrosion products, the galvanized layer was covered with black oxide at the location of the corrosion pits, whose composition was mainly $Fe(OH)_3$, $Fe_2O_3–H_2O$ and $Fe_2(SO_4)_3–8H_2O$. The surface of the steel wire in the uncorroded area around the corrosion pit is still relatively bright, and the degree of corrosion of the steel wire is relatively mild. Through these phenomena, it can be proved that the corrosion type of galvanized steel wire is anodic dissolution type, in which the exposed iron matrix reacts electrochemically with the surrounding corrosion products and H^+ is reduced to atomic hydrogen, and gradually migrates to the deeper layer of the matrix through concentration diffusion (**Figure 3b**).

3. When the galvanized layer is exhausted, the white corrosion product disappears and is replaced by a larger area of red corrosion, and the entire surface of the steel wire is no longer shiny after the corrosion product is removed. The expansion of some corrosion pits becomes slow or even stops expanding. At this point, many corrosion pits begin to fuse with each other, linked together into larger corrosion pits or grooves, whose depth of color is shallow. Due to the merging of small corrosion pits, the number of corrosion pits is significantly reduced (**Figure 3c**).

4. In the late stage of corrosion, red corrosion has completely covered the surface of the steel wire, and after removing the corrosion products, a large amount of iron matrix can be found leaking out, and the surface of the whole steel wire has a naked-eye visible unevenness, and corrosion has occurred in almost all areas to varying degrees (**Figure 3d**).

3.3 Steel wire corrosion fatigue loss of weight

In order to facilitate data processing, the problem of uneven distribution of steel wire density along the axial direction is ignored, and the weight of the steel wire is converted to unit length weight. And use the same method to deal with the corrosion of the steel wire, the data of weight loss over time before and after corrosion are recorded separately (**Table 3**) [14]. The table shows the weight W_1 before corrosion, weight W_2 after corrosion, weightlessness W_3, diameter D_1 after corrosion, average diameter $D_{2,}$ and corrosion time H.

It can be seen from the experimental data, the weight of the steel wire is reduced over time in the experiment, and the growing trend of different stress states of the steel wire is basically the same. In the three stress states, although the weight loss per unit area at a particular moment is not very regular, according to their respective trends to analyze, the overall still follow such a law: In any stage of corrosion of the steel wire, the stress complex steel wire in a corrosive environment is more vulnerable to corrosion.

H (h)	Number	Loading method	W_1 (g/m)	W_2 (g/m)	W_3 (g/m^2)	D_1 (mm)	D_2 (mm)
120	1	Alternating stress	166.8	164.8	125.1	5.23	5.23
	2	Static stress	167.1	165.4	102.2	5.23	
	3	Stress-free	166.7	165.3	86.1	5.23	
240	1	Alternating stress	167.1	163.3	234.1	5.18	5.19
	2	Alternating stress	167.1	163.7	208.2	5.18	
	3	Static stress	167.1	164.8	141.5	5.18	
	4	Static stress	167.0	164.2	171.4	5.20	
	5	Stress-free	167.1	164.8	137.4	5.19	
360	1	Stress-free	166.7	163.3	209.1	5.20	5.20
480	1	Alternating stress	166.9	162.5	271.0	5.17	5.18
	2	Static stress	166.9	162.6	263.2	5.18	
	3	Stress-free	166.6	162.4	257.3	5.18	
840	1	Alternating stress	166.8	160.8	367.5	5.14	5.18
	2	Static stress	167.4	161.9	334.1	5.20	
	3	Stress-free	166.8	161.7	316.8	5.20	
	4	Stress-free	166.9	161.8	313.0	5.16	
1200	1	Alternating stress	167.0	159.9	434.6	5.11	5.16
	2	Static stress	167.2	161.0	379.0	5.14	
	3	Stress-free	167.2	161.6	344.3	5.19	
	4	Stress-free	167.1	161.5	340.6	5.18	
1560	1	Stress-free	166.7	159.3	448.9	5.14	5.13
	2	Stress-free	166.9	159.6	446.3	5.12	
	3	Stress-free	166.6	157.8	539.6	5.12	

Table 3.
Steel wire corrosion fatigue loss of weight date [14].

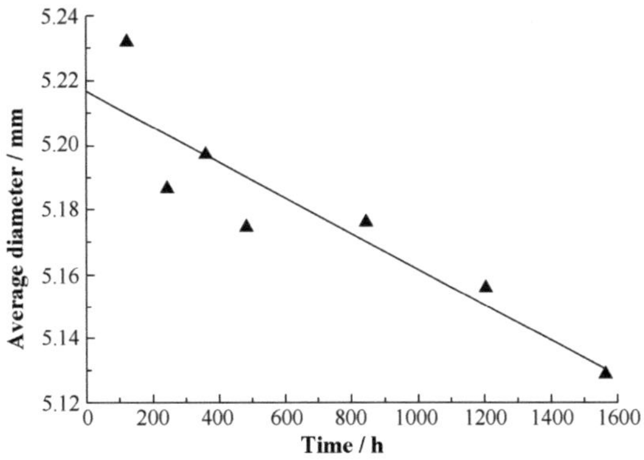

Figure 4.
Relationship between corrosion time and average diameter of each group [13].

Considering the fatigue effect of the wire, alternating stress is not only easy to cause fatigue damage to the steel wire but also may accelerate the chemical corrosion of its surface in a corrosive environment, causing a decrease in the cross-sectional area of steel wire, increasing the stress on the steel wire, and reducing the safety factor of the cable.

Without considering the stress state of the steel wire, the relationship between the diameter of the corroded steel wire and time is plotted in **Figure 4**.

As can be seen from **Figure 4**, the trend of decreasing the diameter of corroded steel wire is linear. The corrosion time is 5 days, and the diameter of the steel wire is 5.23 mm, greater than the standard diameter of the steel wire 5.2 mm, which is due to the fact that the zinc layer thickness will be slightly higher than the specified value in order to protect the quality of the steel wire [13]. The actual experimental wire diameter before corrosion is between 5.23 mm and 5.25 mm, and the effect of 5 days of corrosion on the diameter is not significant, measured with a micrometer will also produce a certain error.

4. Corrosion fatigue damage mechanism and damage evolution model of galvanized parallel steel wire

4.1 Steel wire mesoscopic view analysis based on image grayscale

In order to obtain a quantitative evaluation index, the corrosion degree of the steel wire is evaluated by a gray image-based method to calculate the corrosion pit density of the steel wire, and the degree of corrosion is defined by the percentage of corrosion pit density. Use GSA image analyzer software to perform grayscale analysis of eroded images. GSA Image Analyzer is an image analysis tool that supports all types of 2D images with comprehensive functions to identify and count the number of objects; supports length and area calculation; supports patterns from files, scanners, and microscopes; and can be used in fields such as cell analysis and materials engineering. Since the image formed by the electron beam scanning of the corroded steel wire is a true color image and cannot be recognized directly on the computer, it is digitized and converted into a grayscale image using GSA Image Analyzer. The principle of digital processing is to convert a true-color image into an array of pixels that can be represented numerically, with the value of each dot being the grayscale value of that dot. The spatial resolution of the original image is related to the number of pixel dots there are, the more dots, the higher the pixel resolution is.

With the increase in corrosion time, the local adjacent corrosion pits interlocked, forming larger corrosion pits, and the corrosion pit depth also gradually increased. Under the alternating stress condition and static stress condition, the stress in the pitting area is concentrated and cracks are gradually formed and expanded. The selected corroded wire is cleaned to ensure no residue and then dried in the drying oven. For observation, the dried steel wire was placed on the sample plate to sketch the position, then the sample plate was placed in the sample chamber, the sample chamber was sealed, and the steel wire was observed under the electron microscope. As an example, **Figures 5** and **6** [15, 16] show the microcracking phenomenon on the steel wire surface under alternating stress and static stress conditions.

According to **Figures 5** and **6**, it can be seen that the depth and size of the corrosion pits are gradually developed with the corrosion process. Under the coupling effect of stress and corrosive environment, the nucleation of the corrosion pits of the corroded steel wire is first cracked and forms nanoscale cracks, while the alternating stress makes the nanoscale cracks continuously elongate and retract, resulting in the accelerated formation of microcracks in the steel wire.

As shown in **Figure** 7 [17], the local corrosion map of 300 times under each working condition was imported, and the hue was measured at the parts where the steel wire corrosion was obvious, and the histogram of the hue distribution was derived and the distribution amount was counted to calculate the frequency of corrosion pits. After the numerical processing, combined with the distribution histogram to obtain the corrosion characteristics of the specimen surface parameters can be seen that the corrosion

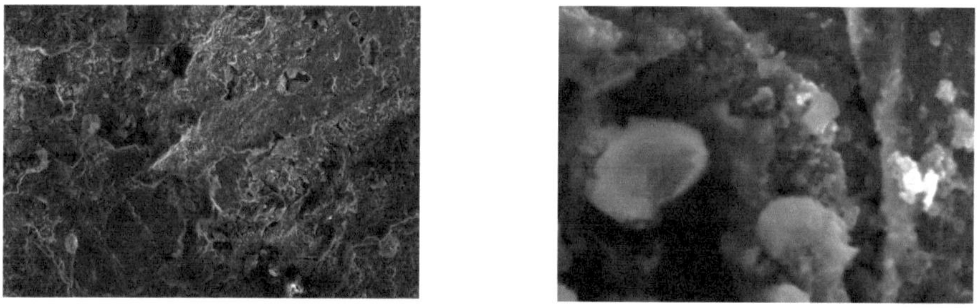

(a) 300 times (b) 6000 times

Figure 5.
Alternating stress microcrack extension [15, 16] (a) 300 times (b) 6000 times.

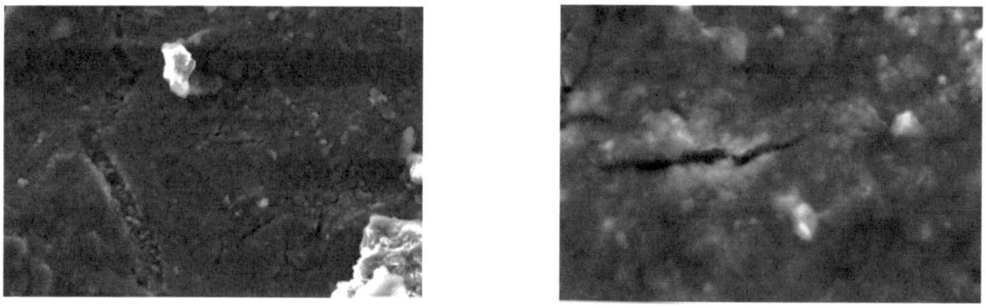

(a) 1380 times (b) 6000 times

Figure 6.
Static stress microcrack extension [15, 16] (a) 1380 times (b) 6000 times.

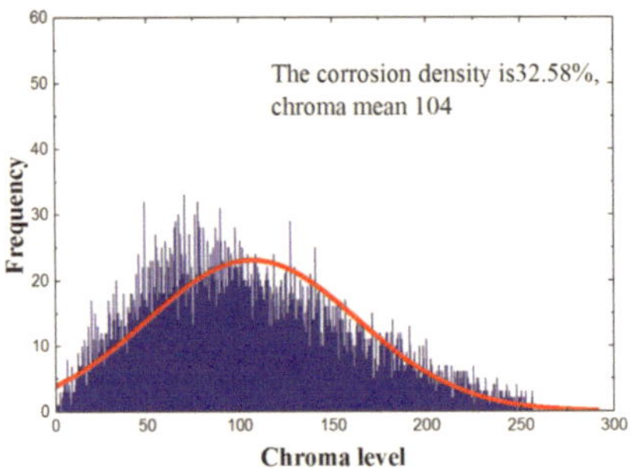

(a) stress-free

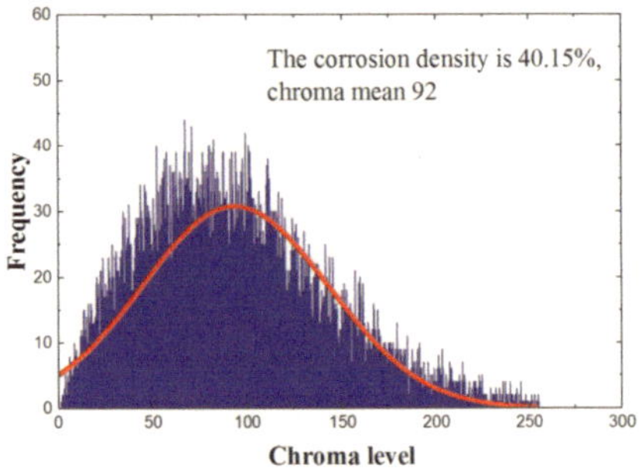

(b) static stress

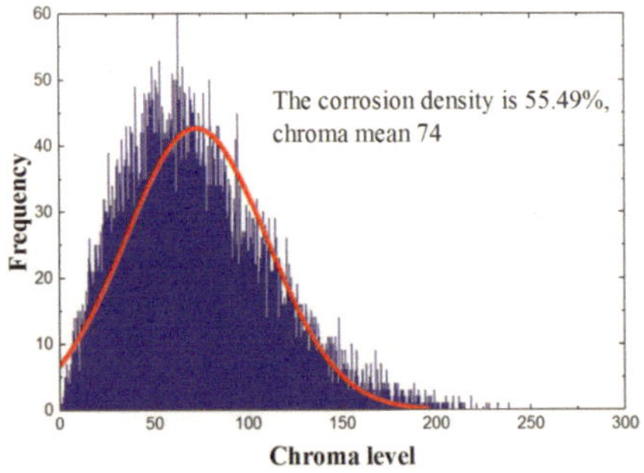

(c) alternating stress

Figure 7.
Corrosion histogram [17] (a) stress-free, (b) static stress, and (c) alternating stress.

characteristics of the steel wire basically conform to the normal distribution. In the salt spray environment under stress-free conditions, there is no obvious corrosion pit in the specimen and the corrosion rate is about 32.58%. The corrosion rate under alternating stress conditions is about 55.49%. Under the static stress condition, the corrosion rate of steel wire is between the previous two conditions, about 40.12%. Wire corrosion is more uniform under stress-free conditions, and the standard deviation of its histogram is larger, with a color tone level of about 104, while the color tone level of the peak gray standard deviation of the static stress condition is about 92, which is between the alternating stress condition and the stress-free condition. For the alternating stress condition, the steel wire surface tone value concentration and corrosion pits with the elongation and retraction of the steel wire to accelerate the development of the peak where the tone level is about 74. Therefore, under the alternating stress condition, the corrosion frequency and corrosion rate of the steel wire are greater than the rest of the conditions, and the damage to the specimen is the largest.

4.2 The mechanical property analysis of corroded steel wire

The corroded steel wire was stretched in one direction on the universal testing machine, and its mechanical property data were recorded, and the broken steel wire of each steel wire was packed and numbered after completion. Select the middle length of the steel wires as a 1-meter section of the experimental section for tensile experiments, the ratio of the maximum load measured before the experimental section is

H (h)	Number	E (GPa)	AVG	R_m (MPa)	AVG	$R_{p0.2}$ (MPa)	AVG	A (%)	AVG
120	1	209.7	212.0	1890	1890	1615	1653	6.5	5.5
	2	210.2		1880		1625		5.5	
	3	216.0		1900		1720		4.5	
240	1	208.2	213.0	1905	1902	1610	1636	5.0	4.6
	2	208.7		1905		1625		2.0	
	3	214.1		1900		1610		4.5	
	4	216.2		1895		1620		5.5	
	5	217.8		1905		1715		6.0	
360	1	213.3	213.3	1900	1900	1730	1730	5.0	5.0
480	1	226.0	229.3	1900	1897	1595	1623	4.5	4.7
	2	303.8		1905		1585		5.0	
	3	232.7		1885		1690		4.5	
840	1	222.6	216.0	1880	1868	1585	1656	5.0	4.5
	2	214.8		1865		1650		4.5	
	3	215.8		1865		1690		4.5	
	4	211.0		1860		1700		4.0	
1200	1	184.8	206.9	1905	1883		1690	2.0	4.0
	2	214.2		1895		1690		5.5	
	3	209.4		1860		1685		4.0	
	4	219.2		1870		1695		4.5	
1560	1	208.8	208.9	1825	1802	1670	1653	3.5	3.0
	2	209.5		1815		1655		3.0	
	3	208.4		1765		1635		2.5	

Table 4.
Mechanical property data of steel wire.

pulled off and its nominal area is the tensile strength, that is, the nominal tensile strength. The mechanical property data are shown in **Table 4**.

According to the analysis of the changes in the mechanical properties data of the steel wire in **Table 4**, it can be seen as follows:

1. The stress state of the steel wire after corrosion is different, its elastic modulus and yield strength vary less, and only the elongation differs significantly through preliminary analysis. Since the corrosion loss weight obtained in the previous article is linearly related to time, the alternating stress, static stress, and stress-free loading state of the specimen elastic modulus with corrosion time change is not significant, the elastic modulus of corrosion is also not sensitive to corrosion performance;

2. In the early stage of corrosion, stress corrosion and corrosion fatigue have little effect on tensile strength, the tensile strength between the three loading states is not significantly different, and the size of the tensile strength has not decreased significantly;

3. In the middle and late stages of corrosion, the tensile strength began to decline sharply. Alternating stress-loaded specimens and static stress-loaded specimens will show a brief increase in tensile strength. This is a hazardous increase and should be given sufficient attention. The decrease in steel wire ductility can be reflected from the side;

4. According to the document in Ref. [18], the post-break elongation of 1860-grade galvanized steel wire shall not be less than 4.0%. And in the late stage of corrosion, the post-break elongation of the three loading states of the steel wire dropped below 4.0%;

5. In the late stage of corrosion, the alternating stress loading causes corrosion fatigue damage to the cable, and its post-break elongation was significantly lower than that of the other two loading modes of the steel wire, which indicates that the alternating load reduced the toughness of the steel wire.

In summary, the mechanical property of the specimens under different loading states was affected to different degrees by the prolongation of corrosion time, and the modulus of elasticity, yield strength, tensile strength, and elongation after fracture all showed a downward trend. Among them, tensile strength and elongation after a break are the most sensitive to corrosion, while the modulus of elasticity is less sensitive to corrosion, tensile strength decreases slowly in the early stage of corrosion, and in the middle and late stages, it decreases in a leap and becomes more sensitive to corrosion, the different loading methods have a greater impact on the yield strength.

4.3 Corroded steel wire fracture morphology analysis

In the static tensile experiment of uncorroded steel wire, it was found that the fracture form is mainly manifested as cup-cone, the fracture form of specimens under stress-free conditions is also a cup-cone fracture (**Figure 8a**), but not as regular as the fracture of uncorroded specimens, the fracture form of specimens under static stress conditions is mainly milling cutter fracture (**Figure 8b**), individual specimen is cup-cone fracture, the fracture form of specimens under alternating stress conditions is mainly cleavage-milled fracture (**Figure 8c**), individual steel wire is miter-type fracture and their combined form, but there are very few cleavage fractures (**Figure 8d**) [15].

From **Figures 8a** and **b**, it can be seen that the cup-cone fracture has an obvious necking phenomenon, the milling cutter fracture also has the appearance of necking but not as obvious as the cup-cone fracture, this necking phenomenon indicates that the steel wire material has obvious plastic properties; while **Figure 8c** and **d**, there is no obvious necking phenomenon in cleavage fracture and cleavage-milled fracture, which shows the brittle properties of steel wire material, and the fracture time and location performance, the fracture time and location are sudden and uncertain. The probability of brittle fracture of the steel wire is the highest under the action of alternating stress and environmental coupling, corrosion fatigue is usually multisource fatigue, the fracture has more fuzzy fatigue striations or brittle fatigue cracks, and there are more secondary cracks in the fracture.

After the mechanical properties experiment on the corroded steel wire, the fracture of the corroded steel wire under alternating stress conditions was scanned with KYKY-2008B digital scanning electron microscope, and the microscopic morphology of the fracture is shown in **Figure 9a–d** [15].

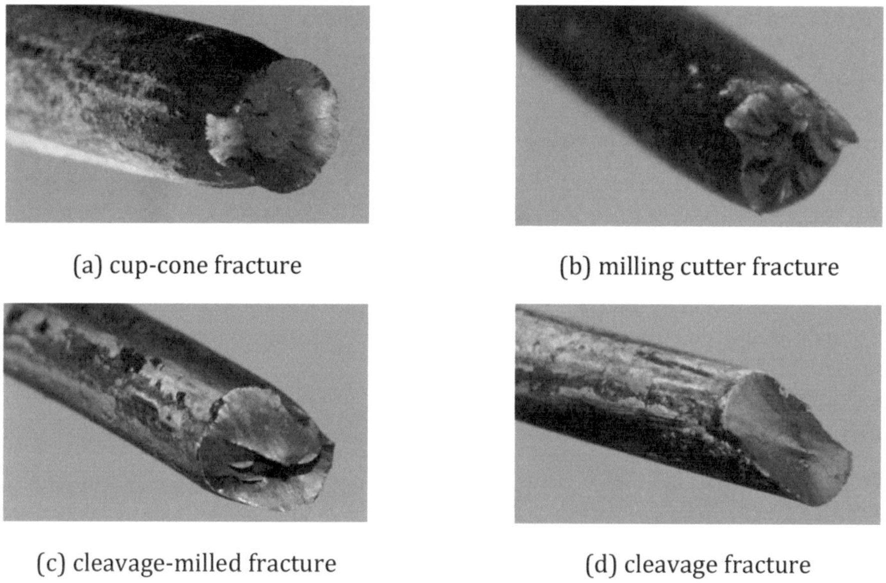

Figure 8.
Fracture morphology of corroded steel wire under different stress conditions [15] (a) cup-cone fracture, (b) milling cutter fracture, and (c) cleavage-milled fracture (d) cleavage fracture.

Figure 9.
Morphology of the fracture under alternating stress [15] (a) 23 times, (b) 340 times, (c) 1700 times, and (d) 5000 times.

After magnification, it can be found that a few sections of the edge of the fracture are neat and show a brittle fracture morphology, while other areas are destructive in shape, and individual areas appear in a river and mountain-like morphology. The center of the fracture appears to have a tough nest-like morphology. When the fracture is magnified to 1700 times by microscope, a "fish-eye" morphology can be seen, and a shell-like fracture appears. This morphology is caused by fatigue, which indicates that the fracture is controlled by the main crack and gradually expands along the expansion of the main crack. As the primary crack grew and expanded, the secondary crack also expanded gradually, which accelerates the fracture of the steel wire. When the fracture is magnified to 5000 times by microscope, a large number of smooth "fish eye" shapes are distributed on the section, according to the shape of the fracture, it can be seen that corrosion of the steel wire is a brittle fracture. The experiment shows that corrosion of severe steel wire plastic properties reduced, brittly enhanced, and the time of fracture is uncertain, with abruptness under the alternating stress and environmental coupling.

5. Working life prediction method and evaluation technology for cable based on breakage safety

5.1 Working life prediction method of service cable

Based on the experimental analysis, compared with non-corroded steel wire, the pitting corrosion, fracture stress, and elongation after fracture of corroded steel wire in the nucleation stage are basically unchanged, the remaining life prediction of this stage is not considered. The process of corrosion pits growing and expanding until they turn into small cracks is a common stage of the steel wire in service cable. At this stage, the life of the steel wire is reduced. After the appearance of the crack, the corrosive environment and the coupling of alternating stresses made the expansion rate of the crack rapidly increase to the maximum, breaking stress and elongation after fracture are sharply reduced, the steel wire toughness is reduced, and brittleness enhanced, there is a turn from tough to brittle at this stage [19]. In order to prevent the appearance of brittle fracture, the steel wire of the cable needs to be avoided in this stage of work, the critical crack appears, and the steel wire brittle fracture will occur at any time, the study of its life is also not considered. Therefore, the reliability of the steel wire of the service cable is affected by the development of corrosion pits until the appearance of small cracks and the appearance of small cracks until the appearance of the critical crack size a_c [15].

The law of fatigue crack expansion can be used to analyze the reliability of in-service ties as well as risk of safety. The fatigue crack expansion rate function commonly used in engineering is the Paris-Erdogan model, that is, the Paris formula, which establishes the relationship between the stress intensity factor and the crack expansion rate and is the theoretical basis for predicting the fatigue crack expansion life in the form shown in Eq. (1):

$$^{da}/_{dN} = C(\Delta K)^n \tag{1}$$

where a is the crack length; N is the number of stress cycles; da/dN is the crack expansion rate; C, n is the material constant, environmental factors such as temperature, humidity, media, loading frequency, etc. are obtained by fitting the experimental data.

ΔK is the stress intensity factor amplitude and is described by Eq. (2):

$$\Delta K = K_{max} - K_{min} = f\Delta\sigma\sqrt{\pi a} \tag{2}$$

where f is a function of member geometry and crack size; K_{max} and K_{min} are the maximum and minimum values of the stress intensity factor at the crack; and $\Delta\sigma$ is the stress amplitude at the crack.

At the stage in which corrosion pits develop into small cracks, each growth of initial defect is independent, the growth of corrosion pit is slow and also affected by accumulation of corrosion product and other factors. while in the small crack until the critical crack size a_c stage, the growth of crack depth is faster, the growth rate increased rapidly to the maximum, the expansion rate is significantly greater than the previous stage.

On the basis of the Paris formula, the corrosion fatigue calculation is simplified into two linear stages, namely, the life of the corrosion pit development to the crack stage N_1 and the life of the tough and brittle expansion stage N_2, the total life of the wire of the cable is equal to the sum of life of these two stages, expansion of corrosion fatigue crack generally meets the deformation form of Paris formula, as shown in Eq. (3):

$$^{da}/_{dN} = D(t)(\Delta K)^m \tag{3}$$

where m is the material constant, generally close to the parameter m in simple fatigue; $D(t)$ is related to the material medium system, loading frequency and load form and other factors, and is a function about time t, instead of the original constant c.

In a comprehensive analysis, the key to applying this method is to determine the range of different stress intensity factors ΔK, the initial corrosion pit size a_0, the initial crack size a_f, and the critical crack size a_c for these two segments. based on this, the life of each stage and the total life N are described by Eqs. (4) to (6).

$$N_1 = \int_{a_0}^{a_f} {da}\Big/\left[\frac{da}{dN}\right]_1 \tag{4}$$

$$N_2 = \int_{a_f}^{a_c} {da}\Big/\left[\frac{da}{dN}\right]_2 \tag{5}$$

$$N = N_1 + N_2 \tag{6}$$

5.2 Safety assessment of damage cable

Cable may fail in two ways after experiencing long-term accumulated damage: One way is that the damage makes the cable reach the critical state that is unsuitable for use and even makes the cable fail under the action of conventional load; another way is that the accumulated damage causes the cable resistance to decay. By now, although the damage does not reach the critical value, part of the steel wire or even the whole cable may suddenly fail when the lasso encounters the action of accidental extreme load. Therefore, it is necessary to effectively detect, monitor, and predict the process of damage accumulation of the cable, grasp the law that its resistance decay with the accumulation of damage, and timely repair and reinforce cable to ensure the safe operation of the cable structure and avoid the occurrence of accidents to the maximum extent possible [19–22].

(1) Evaluation of steel wire corrosion based on appearance: Relying on a unified hard and fast objective standard, so that each tester is bound by this standard and does not rely entirely on subjective guesses to draw conclusions.

(2) Assessment of effective strength reduction factor based on appearance quality: Usually based on the experimental analysis of some representative bridge cables, the empirical formula for regression analysis of cable's strength is used to assess the safety of cables or investigate the influence of defects and diseases of a large number of bridge cables on the bearing capacity of the bridge, determine a more reasonable strength reduction factor according to the investigation results, and then discount the strength of the cable. In this method, the effective resistance of the damaged cable is reduced, and the actual resistance of the damaged cable is equal to the standard value of the steel wire tensile strength of the cable multiplied by the strength reduction factor. The calculation formula is shown in Eq. (7):

$$R^{\tau} = vR^{b} \tag{7}$$

where R^{τ} is the estimated actual tensile strength of the damaged cable, R_b is the design tensile strength value of the cable, and v is the strength reduction factor of the damaged cable.

(3) Safety assessment of cable strength based on the overall level of load effect: A detailed appearance survey is carried out for the cable, and the cable is classified according to the results of appearance survey, and its resistance calculation coefficient is determined according to the classification. The corresponding calculation coefficient Z_1 value is based on a large number of engineering accumulation and experimental statistics. Based on the detailed appearance investigation of the cable, the value of Z_1 can be referred to **Table 5**. The safety assessment of the cable based on the overall level of load effect is to judge the safety condition of the cable by comparing the relationship between the magnitude of the load effect and the resistance of the cable. Z_1 is used as the reduction factor of the resistance of the cable when calculating the resistance of the cable, and the actually allowable resistance of the cable can be calculated by Eq. (8):

$$[R]^{\tau} = Z_1 g[R]^{d} \tag{8}$$

Z_1	Technical status	Status of cable
1.00–1.10	Excellent	Complete surface protection, no water accumulation in the anchor head, and no cracks in the anchorage area.
0.95–1.00	Good	The surface protection is basically intact, with minor cracks, no corrosion of the anchor head, and no cracks in the anchorage area.
0.90–0.95	Moderate	Surface protection with a few cracks, accompanied by a few most corrosion, slight corrosion of the anchor head, and small cracks in the anchorage area.
0.85–0.90	Poor	The protection of surface is generally cracked and partially detached, the anchor head is corroded, and there are obvious stress cracks in the anchorage area.
Below 0.85	Dangerous	The protection of surface is generally cracked and has a lot of shedding, the cable is exposed and rusted severely, and the anchor head is waterlogged and corroded, there are obvious stress cracks in the anchorage area and the width is greater than 0.2 mm

Table 5.
Values of Z_1 discount factor for cable resistance.

where $[R]^d$ is the designed value of the cables' resistance and Z_1 is the reduction factor of the cables' resistance.

(4) Method that evaluates identification factor based on safety factor: For the safety evaluation of cable, the identification factor method from abroad can be used. In the design stage of cable-stayed bridges, a certain safety factor is usually taken for the design resistance of the cable, considering various factors such as material safety, load impact effect, and cable fatigue. The safety factor is defined as the ratio of the load effect to the resistance effect of the cable. Comparing the actual safety factor with the design safety factor, according to the design requirements, if $\eta > [\eta]$, it means that the cable is in a safe state; if $\eta = [\eta]$, it means that the cable is in a safe critical state; if $\eta < [\eta]$, the smaller the measured safety factor η the less safe the cable is under stress. The specific calculation method is shown in Eq. (9):

$$\eta = R/S \tag{9}$$

where η is the safety factor calculated by the actual measurement of the cable, R is the force of cable generated by the load effect, and S is the actual ultimate resistance of the cable.

(5) Assessment methods based on reliability theory: Safety after damage is an important indicator to evaluate the actual working performance of cable-stayed bridges in service. However, due to the differences in the working environment of the structure and the materials themselves, such as geometric characteristics, mechanical properties of the materials, load rating and distribution, and other parameters cannot be unified in the safety assessment of cableways, making the safety assessment of damaged cables uncertain. The assessment method of probabilistic reliability defines this uncertainty as a random variable to achieve the assessment of cable safety. According to the current unified standards for the design and definition of structural reliability of buildings, structural reliability is expressed as the probability that the structure will complete its intended function within a scheduled time and under specified conditions. For the cable, the safety function of the strength is first established as shown in Eq. (10):

$$Z = r - s \tag{10}$$

where s is the ultimate or design strength of the cables and r is the measured or estimated strength of the cables corresponding to s.

Since both r and S are random variables, Z is also a random variable. When $Z > 0$, the structure is in a reliable state; when $Z = 0$, the structure reaches the limit state; and when $Z < 0$, the structure is in a failure state. The static model of reliability analysis is applicable to the resistance of the structure, which hardly changes with time in the process of use, so the resistance can be regarded as a constant value. For real structures, on the other hand, the structural resistance is constantly changing over time, so the resistance of the existing structure must be simulated using a stochastic process. Assuming that the probability density function of Z is $f(Z)$, the general representation of the component safety probability calculation formula is Eq. (11).

$$R = P(Z > 0) = \int_0^\infty f(Z)dZ \tag{11}$$

The key to the safety assessment of cable strength using the reliability method is to establish the functional function and determine the distribution parameter.

The general functional functions are divided into two types, one is based on the functional function of cable resistance and another is based on the safety factor. The structural resistance and load effects in the functional functions generally obey normal distribution, lognormal distribution, and Weibull distribution. The relevant research results suggest that for generally engineering design, it is reasonable and biased toward safety to use normal distribution for member resistance and load effects. For members with high safety requirements, lognormal or two-parameter Weibull distributions are recommended.

(6) Fatigue-based safety assessment of cable: For the safety analysis of fatigue strength of cable, the Palmgren-Miner linear cumulative damage theory is commonly used in engineering, and the corresponding stress-fatigue life curve is statistically analyzed and summarized through the steel wire fatigue experiment, form a standard for the service life of cables under different load cycles, which is used to predict and evaluate the fatigue safety performance of cable.

6. Conclusions

In this chapter, through the corrosion fatigue experiment on the steel wire of bridge cable under the action of coupled environment and load explore the fine morphology of the steel wire under different corrosion time and different stress loading, analyze the law of evolution of mechanical property and the different fracture morphology of the steel wire, and analyze and generalize the service life and damage assessment technology of the cable based on the breakage safety theory, the main conclusions are as follows.

As the steel wire of the cable undergoes electrochemical reaction during corrosion, the surface of the steel wire produces white corrosion products. As the corrosion deepens, the galvanized layer on the surface of the steel wire is depleted, the white corrosion products disappear and are replaced by a larger area of red corrosion, and the surface of the entire steel wire appears uneven as visible to the naked eye.

The corrosion characteristics of the steel wire are basically in line with the normal distribution, in the salt spray environment, the steel wire did not appear obvious corrosion pits under stress-free conditions, the corrosion rate is about 32.58%; the corrosion rate of alternating stress conditions under alternating stress conditions is about 55.49%; Under the static stress, the corrosion rate of steel wire is somewhere in between, about 40.12%. Under stress-free condition, steel wire corrosion is more uniform, and the frequency of steel wire corrosion and corrosion rate is greater than the rest of the conditions, the degree of corrosion performance: Stress-free < static stress < alternating stress under stress-free condition.

Corrosion in the early stage, stress corrosion, and corrosion fatigue have little influence on tensile strength. In the middle and late stages of corrosion, the tensile strength began to decline sharply, the tensile strength of the steel wire appears briefly increased phenomenon under alternating stress and static stress conditions, and the ductility of steel wire decreased. In the late stages of corrosion, the alternating stress causes fatigue damage to the steel wire during the corrosion process, the toughness of the steel wire is significantly reduced, elongation after fracture of the steel wire is significantly lower than the other two loading states, at this time, the possibility of brittle fracture of the steel wire is high.

The stress of the steel wire bundle of the cable is in a dynamic redistribution process with the development of corrosion, and the cable will fracture rapidly when the tensile force reaches the maximum bearing capacity. In order to ensure the safety of the cable in service, it is necessary to predict the crack expansion and the remaining service life of the cable in service based on the linear-elastic fracture mechanics, and to evaluate the safety performance of the existing damaged cable using different methods according to the actual project.

Acknowledgements

This chapter was supported by the National Natural Science Foundation of China (Grant No. 52178273), the Natural Science Foundation of Chongqing (Grant No. cstc2021jcyj-msxmX1159), the Chongqing Talent Plan Project (Grant No. cstc2022-ycjh-bgzxm0124), the Open Fund Project of State Key Laboratory of Mountain Bridge and Tunnel Engineering (Grant No. SKLBT-YF2105), the Chongqing Project of Joint Training Base Construction for Postgraduates (Grant No. JDLHPYJD2020004) and the Innovation Program for Graduate Student of Chongqing (Grant No. CYS22392 and 2021S0004).

Author details

Guowen Yao[1,2*], Xuanbo He[1,2], Jiawei Liu[1,2], Jiangshan Lu[1,2] and Zengwei Guo[1,2]

1 School of Civil Engineering, Chongqing Jiaotong University, Chongqing, China

2 State Key Laboratory of Mountain Bridge and Tunnel Engineering, Chongqing Jiaotong University, Chongqing, China

*Address all correspondence to: yaoguowen@sina.com

References

[1] Yanaka Y, Kitagawa M. Maintenance of steel bridges on Honshu-shikoku crossing. Journal of Constructional Steel Research. 1982;**58**(1):131-150

[2] Barton SC, Carson CC, Gill RS. Implementation of pyrolysate analysis of materials employing tagging compounds to locate an overheated area in a generator. IEEE Transactions on Power Apparatus and Systems. 1982;**PER-1**(12): 4983-4989

[3] Kheyroddin A, Saghafi MH, Safakhah S. Strengthening of historical masonry buildings with fiber reinforced polymers (FRP). Advanced Materials Research. 2010;**133–134**:903-910

[4] Dolley EJ, Lee B, Wei RP. The effect of pitting corrosion on fatigue life. Fatigue and Fracture of Engineering Materials and Structures. 2000;**23**(7): 555-560

[5] Mahmoud KM. Fracture strength of a high strength steel bridge cable wire with a surface crack. Theoretical and Applied Fracture Mechanics. 2007; **48**(2):152-160

[6] Widyawati R, Takahashi J, Emoto H, Miyamoto A. Remaining life prediction of an aged bridge based on concrete core test. Procedia Engineering. 2014;**95**:88-99

[7] Mahmoud KM, Fisher JW. Mechanics of environment-assisted cracking in bridge cable wire. Bridge Structures. 2007;**3**:3-4

[8] Ou L, Gang S, Bo W, Chengxin Z. Inspection of technical condition of bridge stay cable in operation period. World Bridges. 2017;**45**(04):79-83

[9] Jun X, Weizhen C. Detection and analysis of the cable deterioration of Shimen bridge. Steel Construction. 2007; **05**:81-84

[10] Hailiang W, Jiafei X. Analysis of the cause of a bridge cable damage and suggestions. Railway Standard Design. 2005;**10**:67-69

[11] China Quality Inspection Press. Corrosion Tests in Artificial Atmospheres—Salt Spray Tests (GB/ T10125-1997). Beijing: China Quality Inspection Press; 1997

[12] Dongxia J. Experimental Study of Cable Environmental Corrosion Damage under Alternating Stress. Chongqing: Chongqing Jiaotong University; 2013

[13] Guowen Y, Shiya L, Zengwei G. Corrosion Fatigue Performance and Damage Mechanism of Bridge Cable Structures. Beijing: Science Press; 2022

[14] Li Z. Damage and Safety Performance Evaluation of the Cable-stayed Corrosion. Chongqing: Chongqing Jiaotong University; 2013

[15] Shicong Y. Research on the Corrosion-Fatigue Problems and Service Reliability of the Bridge Cables and Hangers. Chongqing: Chongqing Jiaotong University; 2018

[16] Shicong Y, Jinquan Z, Guowen Y. Microscopic damage behavior of corroded steel strands based on image gray analysis. Chinese Journal of Solid Mechanics. 2018;**39**(03):305-315

[17] Shicong Y, Guowen Y, Jinquan Z. Observations on the damage behaviors of corrosion fatigue in steel strands based on image analysis. Advances in Mechanical Engineering. 2017;**9**(12): 1-10

[18] China Standard Press. Hot-dip Galvanized Steel Wire for Bridge Cable (GB/T 17101-2008). Journal of Chongqing Jiaotong University (Natural Science). Beijing: China Standard Press; 2008

[19] Guowen Y, Chaoyue L, Guoqiang W. Mechanism of corrosion damage of stayed cable under the effect of acid rain and loading coupling. 2016;**35**(06):6-10

[20] Guowen Y, Xuesong C, Li Z. Study on corroded cable evaluation based on gray image analysis. Journal of Chongqing Jiaotong University (Natural Science). 2016;**35**(04):10-12

[21] Wenjie W. Experimental Study of Stay Cable Electrochemical Corrosion Fatigue under Alternating Stress State. Chongqing: Chongqing Jiaotong University; 2015

[22] Guoqiang W. Experimental study on Stress Corrosion and Corrosion Fatigue of Stay Cables under Acid Rain Condition. Chongqing: Chongqing Jiaotong University; 2015

Chapter 7

Erosion-corrosion

Sajjad Akramian Zadeh

Abstract

Generally, almost all components moving near a corrosive fluid hitting the material surface are exposed to corrosive erosions. Meanwhile, transmission pipes of gas, oil, and water, the transmission lines of corrosion fluid in the industrial reactor, and heat exchange systems are suffering significantly from the erosion-corrosion phenomenon. Erosion-corrosion can generate material loss much greater than the sum of the pure erosion and the pure corrosion individually due to the interaction between them. Erosion-corrosion in aqueous systems is dominated by two major mechanisms: electrochemical corrosion and mechanical erosion. On account of the greater material loss than the sum of their components, the interaction between electrochemical and mechanical processes has been recognized in many works, and they have been referred to as "Synergistic" and "Additive" effects. The so-called synergistic effect is normally used to describe how corrosion can enhance erosion, while the so-called additive effect refers to the mechanism by which erosion can enhance corrosion. In general, the influencing parameters in this process include: the solid sand particles (mass, hardness, density, size, shape, velocity, and impact angle), target material (hardness, metallographic structure, strength, ductility, and toughness), and the environment (slurry composition, flow velocity, and temperature).

Keywords: erosion-corrosion, erosion, corrosion, fluid velocity, corrosive media, corrosive wear

1. Introduction

Every year, erosion-corrosion causes severe damage in slurry transportation and related industries [1]. Many groups from the industry and universities have spent a lot of effort to understand this phenomenon and ways to prevent or reduce its effects. In erosion-corrosion, there are three separate phenomena: impact of solid particles (on the surface of the material), electrochemical reactions, and fluidity of the medium. These processes interact with each other and create erosive corrosion, forming a completely complex phenomenon. **Figure 1** shows the interaction between erosion, corrosion, and the fluid environment. This shows that the erosive corrosion process can only exist if erosion, corrosion, and the fluid environment exist at the same time.

1.1 Separate processes in erosion-corrosion

Erosion processes include cutting, plastic deformation, or contact fatigue caused by cavitation from fluid flow or solid particle collisions. Therefore, erosion is

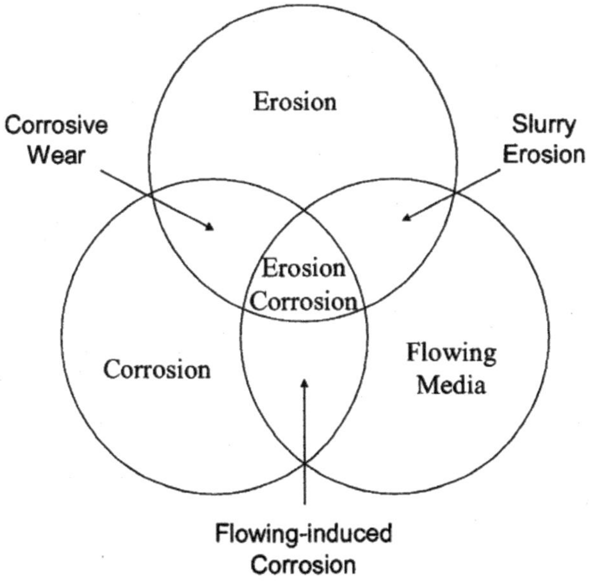

Figure 1.
Schematic representation of the interaction of erosive corrosion phenomena [2].

essentially a process of mechanical loss of material. The process of corrosion is the decomposition of atoms of matter by an anodic current. Atoms of matter in solution are decomposed into their ionic state. Therefore, corrosion has the essence of an electrochemical phenomenon. Together, these two processes can create an additional effect that leads to a complex problem. Therefore, to study this problem, it is necessary to understand each of the separate phenomena in erosive corrosion.

1.2 Pure corrosion

Bradford [3] defined corrosion as "damage caused to a metal due to reaction with the environment." Corrosion always occurs at the surface of the metal where it is in contact with the environment, such as soil, a solution, or even moist air. From a thermodynamic point of view, most metals are unstable when in contact with the environment. They tend to lose electrons and become more stable, that is, cations, oxides, or other chemical compounds, which is the opposite of extractive metallurgy. As mentioned above, there is charge transfer in corrosion processes; hence, this process is considered electrochemical. When corrosion occurs, metal atoms are absorbed into the solution. In most cases, there are three basic steps to this process: (a) transfer of reactant to the electrode surface, (b) a surface electrochemical reaction, (c) transfer of products away from the electrode surface. Each of these steps has its complexity [4].

1.2.1 Two differences in corrosion systems

The corrosion system can be divided into two major parts based on what becomes a protective layer on the surface. One is the active or non-passivated system, and the other is the passive system. In active systems, the material surface is not protected by an oxide layer. In these systems, uniform corrosion always occurs. When the metal atoms react with the electrolyte, some products may form on the surface. Typically, these products are neither dense nor protective. This shows that the metal atoms can dissolve in the electrolyte without any significant hindrance. Galvanic corrosion, intergranular corrosion, and crevice corrosion occur in these active systems. In these systems, the controlling step is charge transfer on the metal surface or resistance to mass transfer in the fluid environment or both.

In passive systems, the surface of the material always becomes a protective oxide layer. This protective film can be either a surface absorption film or a three-dimensional oxide film. The passive film can reduce the corrosion rate by preventing the material from contacting the electrolyte or by limiting the movement of ions and atoms. Therefore, the structure and characteristics of the passive film can affect the corrosion behavior of the material. Passive systems always have a lower corrosion rate than active systems. However, in these systems, localized corrosion always occurs and causes severe damage, such as pitting corrosion. For example, in pitting corrosion, holes are created in the pitted area, but the rest of the area is still covered by the protective layer, so it is not corroded. This is a typical system with a small anode and a large cathode, so the rate of anode dissolution is significantly increased [4]. Over time, the holes become deeper and deeper and eventually penetrate the entire material.

1.2.2 Corrosion in fluid conditions

In a corrosive environment, the current affects the corrosion rate. A fluid medium increases mass transfer, thus increasing the overall corrosion rate. The presence of a fluid medium increases the corrosion rate when the material is subject to general or uniform corrosion. At high speeds, the current can delay the formation of protective films or the passivation process. On the other hand, metal materials themselves suffer from mechanical damage at appropriate high speeds [5, 6].

1.3 Erosion

Erosion is a special state of wear in which the process of tearing off pieces from the surface of the part and thinning due to interfacial tensions applied by a fluid (liquid or gas) or collision of suspended particles in the fluid occurs [7]. A type of erosion called spark erosion is caused by the impact of sparks or an electric arc on the surface of the material, which is used in the industry for machining materials. Another type of erosion occurs at very high speeds called “ultrasonic effects.” In the present discussion, erosion means erosion caused by mechanical factors in common industry conditions. The pure mechanical erosion rate (ER) depends on the fluid velocity as follows [8]:

$$ER\left(\frac{mm}{year}\right) = K_m K_{En}.c.v^n.f(\beta) \qquad (1)$$

In this formula, K_m is the material factor (depending on hardness and flexibility), K_{En} is the environmental factor (including size, shape, hardness, and density of particles suspended in the fluid), c is the concentration of particles, n is the power of velocity, v is the velocity of particles, and β is the angle of impact of particles on the surface. Are. The value of n is about 3 in most cases. The effect of the impact angle (β) on the erosion rate depends on the type of material (brittle or flexible). This issue is shown in **Figure 2**.

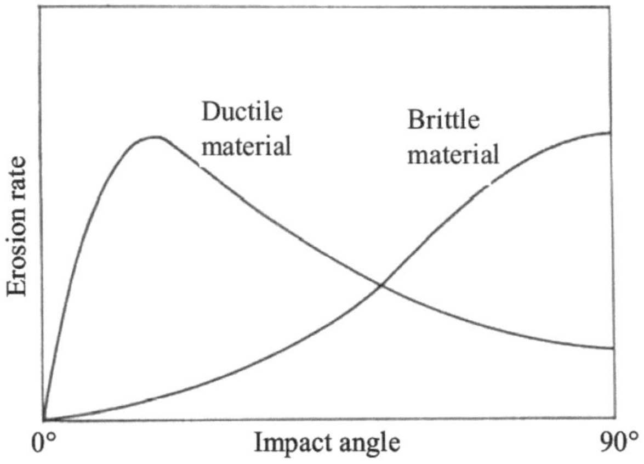

Figure 2.
Effect of impact angle on the rate of erosion of soft and brittle materials [8].

2. Synergistic effect

The rate of erosion-corrosion is higher than the sum of the rates of pure erosion and pure corrosion separately. The additional rate of material loss is caused by the interaction between corrosion and erosion. Many researchers have conducted extensive studies based on this interaction. Matsumura [9] reported that in a NaOH slurry, the loss of pure iron in the presence of corrosion flow was higher than that without corrosion flow, although the corrosion flow could be reduced to a very small extent. It was a measurement. Li and his colleagues [10] investigated the loss of aluminum in an acidic slurry containing 0.5 M NaCl and found that 40–50% of the total weight loss is related to the synergistic effect. Yui and his colleagues [11] reported that up to 86% of the total weight loss of white iron and chromium (chromate) in a slurry with low pH can indicate the synergistic effect. Zhang et al. [12] found that the weight loss due to the synergistic effect can be more than 92.1% for X60 tube steel, 94.6% for 321 stainless steel, and even 99.8% for 316 L stainless steel. Therefore, the important effect of the "synergistic effect" should be considered according to the above studies.

Although the erosion-corrosion map was made by Lim and Ashby [13], it can only quantitatively determine the erosion rate, corrosion rate, and ratio between them. It cannot explain how erosion and corrosion interact with each other or the mechanism of their effect. Many efforts have been made to investigate the mechanism of erosion-corrosion, but due to its specific complexity, it has not yet been fully determined. Researchers mainly divide the "synergistic effect" into two groups: erosion affecting corrosion and corrosion affecting erosion or corrosion increased by erosion and vice versa.

2.1 The effect of erosion on corrosion

As mentioned above, the effects of the erosion process include fluidity, corrosive environment, and particle impact. In a passive system, these effects lead to surface roughness, increased mass transfer acceleration, passive film failure, and increased localized corrosion. In an active system, these can lead to all of the above results except the failure of the passive layer, because there is no passive layer in the active system.

2.1.1 Surface roughness

The impact of particles makes the surface of the electrode rough and uneven. Normally, the rougher the surface, the higher the corrosion rate [14, 15]. When the electrode surface is hit by the particles, an impact cavity can form a crack or become sharp. Li and his colleagues [14, 15] used copper as a working electrode and researched the copper electrode. They found that it is easier for the electron to escape from the surface in the state of a sharp scratch than in the state of a gap. This shows that corrosion is easier in the pointed scratch mode than in the crevice and valley mode. For a completely rough surface, there is higher electron mobility and more freedom to react with the environment, so the surface is more electrochemically active. In addition, in the form of a sharp scratch, there is a tendency to lose more electrons and a more active surface. In contrast, the gap is nobler. This states that a galvanic couple can be formed, meaning that the pointed scratch acts as the anode, and the gaps act as the anode. Therefore, the localized cell formed can accelerate the corrosion process.

Burstein and his colleagues [16] investigated the relationship between surface roughness and pitting potential for 304 L stainless steel and found that surface condition is a critical parameter for determining pitting potential. They found that surface roughness caused by erosion has less pitting potential. He also noticed that during the erosion process, the potential of pitting decreases more than after the process. This shows that in addition to surface roughness, other parameters affect the pitting potential.

2.1.2 Failure of the passive layer

As mentioned, for a passive system, the surface of the material is covered by an oxide film that prevents the collision between the surface of the material and the environment. As a result of the erosion process, the surface of the passive layer may be particle collision or pitting will be damaged. Li et al. [17] and Zhao et al. [18] used the scratch test to simulate the impact of solid particles and obtained similar results to those obtained from the impact of solid particles. Some researchers [19, 20] reported that if the slurry concentration or velocity is high enough, it can replace the corrosion behavior; it means removing the passive area.

Failure of the passive film can increase pitting corrosion. When a particle hits the passive surface, its abrasion removes the passive surface and produces small holes. Burstein et al. [21] investigated the same effect on stainless steel and found that semi-stable pits can form after a short time below the rapid pitting potential. Also, it can increase the recurrence of semi-stable cavities compared with free erosion conditions. These small holes can shorten the period of pitting corrosion. The continued collision of particles ends the expansion of the small holes that created the large holes.

2.1.3 Acceleration of mass transfer

For a corrosion process, there are always two reactions: a cathodic reaction and an anodic reaction. When the electrolyte is fluid, it increases the transport of oxygen, the speed of the cathodic reaction increases, and as a result, the entire corrosion process improves. Also, the fluidity of solid particles in a solution leads to a disturbance in the fluid. Therefore, it can increase the transport processes of both reactants and corrosion products [12].

2.1.4 Strain-hardening effect

When particles collide with the material surface with a high and sufficient kinetic energy, they can cause both elastic and plastic deformations on the surface. This process includes a deforming effect, such as a misplaced lattice, deforming bands, or the protrusion of slip plates on the surface. Li and his colleagues [22] investigated the relationship between the effect of strain rate and electron work performance and found that a higher strain rate makes it easier for electrons to escape from the surface of the material and increases the tendency to corrosion. Li and his colleagues [23] also investigated the change of current density in stainless steel, caused by (changing) the strain rate in the solution $\%10H_2SO_4$, and found that the current density increases with the increase of the strain rate. Mayozumi and his colleagues [24] found that cold working can affect the corrosion behavior of 304 stainless steel.

However, there are some conflicting results on the effect of strain and cold work. Lu et al. investigated the effect of strain on 304 stainless steel and found that in the quasi-elastic deformation process, neither elastic nor plastic deformation has significantly changed the corrosion rate of the electrode when it undergoes an anodic decomposition process. Mayozumi and his colleagues [24] also found that the effect of strain on corrosion is significantly dependent on the system. In Li's research [23], it is difficult to say that the failure of stainless steel passive film was due to tensile stress or strain rate that increased the current.

2.2 The effect of corrosion on erosion

Corrosion is a chemical process, specifically the breakdown of a surface layer of material. When solid particles hit the surface, the erosion process is much easier than when there is no corrosion flow at all, and this leads to a decrease in erosion resistance in the surface layer of the material [25, 26]. Processes such as roughening of the surface, preferential decomposition of the background phase, removal of the hardened surface, generation of vacancies, and chemical mechanical effect are studied in this section.

2.2.1 Surface roughness

When the microstructure of the material is not uniform, the decomposition process on the surface of these materials will not be uniform. Some areas are more susceptible to corrosion while others are not. As a conclusion, the more active areas act as the anode and the nobler areas act as the cathode. Therefore, we see metal decomposition in the anode. While this is not the case in the cathode. In these conditions, unevenness is created. Postlethwaite [27] investigated the erosive corrosion behavior of pipes in an aqueous slurry, with and without the presence of inhibitors. Inhibitors can slow down the corrosion process. He found that the rate of erosive corrosion and surface roughness can be reduced by adding inhibitors. Some researchers [28] explained that, because the erosion process is sensitive to the angle of impact of particles on the surface, corrosion by roughening the surface can change the angle of impact.

2.2.2 Preferential decomposition of the matrix phase

For some materials, the mechanical properties are enhanced by a secondary dispersed phase in the metal matrix composite (MMC). For example, the reinforced phase can increase the surface hardness of the (MMC) and can make the material more resistant to wear and friction. The friction of the surfaces in contact is more resistant. For these materials, the mechanical properties are much more dependent on the reinforcing phases. When these materials are exposed to erosive corrosion, preferential decomposition always occurs at the junction of the ground and reinforcing phase, because it is more galvanically active; hence, they are more susceptible to corrosion [29]. This preferential degradation weakens the composite bonds and the metal substrate. Therefore, the reinforced composite may be destroyed earlier by the abrasive particles. This effect can reduce the mechanical properties of the surface of the material; as a result, the erosion resistance of the material decreases [30].

2.2.2.1 Removal of hardened surface

Some materials have the ability to increase hardening. When the erosion process occurs, the abrasive particles continuously hit the surface, and this causes a cold working to be induced on the surface of the material. This cold working induces a hardening effect on the surface and protects the material from further loss by erosion or, in other words, reduces the erosion process. Matsumura and his colleagues [9] investigated the erosive corrosion process in a passive system and found that particles cause the wear of the passive protective film and accelerate the corrosion process. This corrosion flow decomposes the hardened layer on the surface of the material; hence, the erosion resistance decreases.

2.2.2.2 Creating a cavity on the surface layer

Preferential decomposition theory can explain the decrease in mechanical properties of composite materials, but it cannot be used to explain the decrease in mechanical properties of single-phase materials. Some researchers also researched the erosion-corrosion process of single-phase materials. Zhuo and his colleagues [31] investigated the reduction of mechanical properties of pure iron under anodic decomposition conditions in solution by *in situ* nano dentations and found that the surface hardness significantly decreased in the presence of a corrosion current compared with the conditions under cathodic protection. Matsumura et al. [9] reported that in a NaOH slurry, material loss from pure iron in the presence of a corrosion current was 20% higher than in the case without corrosion, although the corrosion current was measured to be very small. Jones et al. [32] proposed the vacancy generation theory to explain this effect for single-phase material. When a material is exposed to anodic decomposition, a supersaturated state of vacancies is formed on the surface of the material by anodic decomposition. A large amount of vacancies leads to the weakening of interatomic bonds on the surface of the metal. The interatomic bonds create a chemical potential gradient between the surface layer and the bulk material, which causes the induced decomposition of anodic and dislocation movement in the surface layer. During the electrochemical decomposition, the vacancies are attracted to the dislocations and increase their kinetic energy. This increase in the kinetic energy of the dislocation reduces the resistance of the surface layer against the change of plastic shape. Others [33] also found that the weakening of interatomic bonds on the surface of the metal leads to the deterioration of the mechanical properties of the material, such as tensile strength, modulus, and fatigue life; hence, it causes a decrease in erosion resistance.

Vacancies also cause corrosion and a chemical mechanical effect. A chemical mechanical effect occurs when a load is applied to the surface of a material while a chemical reaction is taking place. As a result, the mechanical properties are affected by this process. Gutman [34] found that on the surface of a metal, electrochemical or chemical decomposition can lead to a decrease in free energy and an increase in the kinetic power of the dislocations produced in this process. Therefore, the resistance to plastic deformation decreases. Jones et al. [32], and Zhu et al. [35] suggested that anodic decomposition can create a supersaturation of vacancies on the metal surface. These voids penetrate the grain boundaries and isolated voids, causing acceleration of infiltration creep and the ascent of dislocations.

2.3 Factors affecting the erosion-corrosion process

As mentioned above, erosion-corrosion is defined as the interaction of solid particles, fluid medium, and corrosion. This process is considered an interdisciplinary study between materials, hydrodynamics, and electrochemical properties. Therefore, the factors that affect each of these processes also affect erosion-corrosion.

2.3.1 Material properties

2.3.1.1 Microstructure

Resistance to erosion-corrosion is almost dependent on phase composition and particle size [36–39]. Wang et al. [36] investigated the erosion-corrosion behavior of carbon steel and found that the lower bainite microstructure can increase the erosion-corrosion resistance of carbon steel. Patterson and his colleagues [37] investigated the effect of different microstructures on carbon steels in most of the impact angles and found that their erosion resistance is based on the increase of erosion speed in the following order:

Spherodized, pearlite, tempered martensite, and martensite.

Lindsley et al. [38] reported that in spheroidized Fe-C alloys, the wear resistance increased when the mean path between carbides and grain boundaries decreased. Berglozzi [39] reported that erosion resistance is also dependent on the particle size and the erosion resistance increased with decreasing particle size.

2.3.1.2 Composition

The resistance of metals and alloys against erosion-corrosion depends on their chemical composition, corrosion resistance, hardness, and metallurgical history. The corrosion resistance of metals and alloys is mainly determined by the chemical composition. If it is an active metal, or an alloy that consists of active elements, its corrosion resistance mainly depends on the ability to form and maintains a protective shell. If the metal is nobler, it has good inherent corrosion resistance. Therefore, if all other factors are equal, a metal with higher intrinsic resistance will be more resistant to erosion-corrosion [40].

In general, the increase of carbon in carbon steels increases the resistance to erosion, but the resistance to corrosion decreases [28]. For alloy steels, the addition of alloying elements such as Ni, Mn, Mo, and Cr improves the corrosion resistance of the alloy. Samy [41] reported that alloy steels with higher Cr have better erosion-corrosion resistance because increasing Cr increases both corrosion resistance and mechanical properties.

2.3.1.3 Surface shells

The nature and properties of surface protective shells that are formed on some metals and alloys are very important in terms of resistance to erosion-corrosion. The ability of these shells to protect the metal depends on quickly or easily forming them in the early stages of contact with the corrosive environment, their resistance against mechanical damage or erosion, and the speed of their re-formation in case of damage or destruction. A surface shell that is hard, dense, sticky, and continuous is better protection than when the shell is easily worn or peeled off. If the shell is brittle and cracks and crumbles under stress, it will no longer be protective. Sometimes, the nature of the protective shell that forms on the surface of the metal will depend on the corrosive environment in which the metal is located, which is a determining factor.

The changes in the corrosion rate of steel by static water at different pH are dependent on the nature and composition of the surface shells that are formed. **Figure 3** shows the effect of the pH of distilled water at 50°C on the erosion-corrosion of carbon steel. Corrosion speed is low at pH 6 and 10, and corrosion speed is high at pH 8 and less than 6. At pH less than 5, the shell cracks, which is probably due to internal stresses, and the surface of the metal is exposed to the environment. In the areas where the corrosion rate is low, the corrosion products are Fe(OH)2 and Fe(OH) 3, which are more protective because they have prevented the transfer of oxygen and

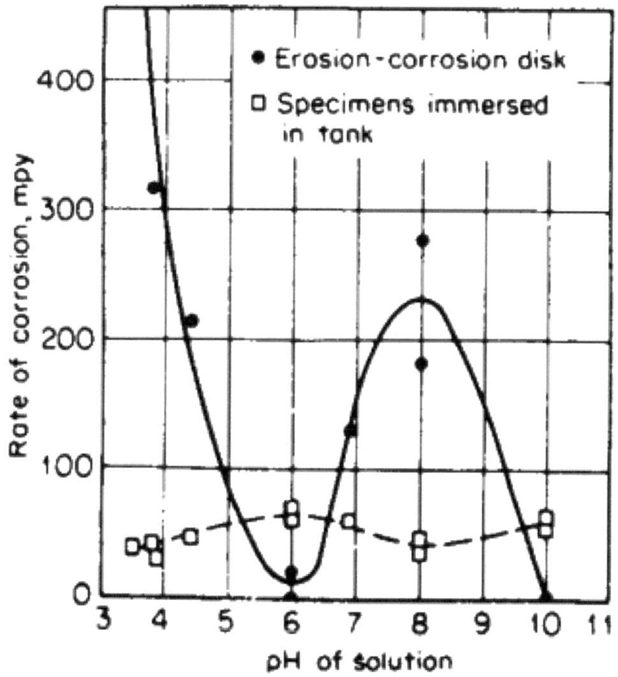

Figure 3.
Effect of PH of distilled water on erosion-corrosion of carbon steel at 50°F (velocity 39 ft./sec) [40].

ions. Erosion-corrosion tests in a boiler at 250°F using different equipment, as well as the experiences of power plants, confirm the results of high corrosion rate at pH = 8. The behavior of steel pipes and low-alloy steels against petroleum materials at high temperatures in refineries depends on the sulfide shell formed on the surface of the pipes. When the shell is worn, high corrosion occurs. For example, in organic systems, if cyanides are present, the sulfide shell, which is solid and integrated, became porous and will no longer be protected. The correct and effective use of inhibitors in reducing erosive corrosion in most cases depends on the nature and type of shell on the metal and, as a result, the reaction between the metal and the inhibitor [40].

2.3.1.4 Mechanical properties

Mechanical properties such as hardness, toughness, strength, and strain hardness can influence the rate of erosion-corrosion by changing the erosion behavior. Many erosion models are mainly created by eliminating these mechanical features. Hardness is one of the important parameters that has attracted many researchers in the field of erosion-corrosion. It is generally said that increasing hardness leads to decreasing erosion. Many researchers who built erosion models have all shown that the erosion rate has an inverse relationship with the hardness of the material [42]. The following equation was created in this regard:

$$e_{\circ} = k_H H_v^{-n_H} \tag{2}$$

That k_H and n_H are laboratory constants that are related to hydrodynamic conditions. H_v is the hardness of the bulk material. Oka and his colleagues [43] found that it is not the initial hardness of the material that affects the erosion process, but the hardness that affects it during the erosion process, that is, the effect of hardening. Hutchings [28] investigated the erosion behavior of a wide range of materials and found that the erosion rate is strongly dependent on the hardness difference between the target material and the impacting particles. Finnie [44] also investigated a range of materials and found that the erosion rate is inversely proportional to the hardness if the power constant is equal to 1.

When there is corrosion, the mechanical properties of the material change. Guo [31] and his colleagues investigated the degradation of the mechanical properties of the pure iron surface in a solution with current density of 0.5 mA/(cm^2) and 1 mA/(cm^2) and found that the mechanical properties of corrosion flow are reduced. Based on this, it can be concluded that the reduction of erosion resistance in the erosion-corrosion process is probably due to the loss of the surface mechanical properties of the material caused by the corrosion flow. Gottman [34] also performed the necessary experiments and theoretical analysis and found that the hardness and strength of the surface layer decrease with the increase in the density of the anodic decomposition current. Lu and Liu [45] have shown a theoretical prediction of the relationship between the increased erosion rate with the presence of a corrosion current and surface degradation. This relationship is as follows:

$$\frac{e_c}{e_{\circ}} = -n\frac{\Delta H_v}{H_v} \tag{3}$$

Where ΔH_v is the difference between the hardness in the corrosion process and the hardness without the corrosion process, and n is the laboratory constant, which is the difference between the erosion systems.

2.3.2 Hydrodynamic properties

2.3.2.1 Flow rate

The slurry is a two-phase fluid system including liquid and solid. Contrary to a single-phase system, the collision of particles should be considered. Meng and his colleagues [46] found that the speed of particle collision is an important parameter in determining the wear rate, and this parameter is directly dependent on the speed of the slurry. Postlewaite and his colleagues [47] have shown the relationship between the erosion-corrosion rate and the flow rate in Eq. (4):

$$w = m_1 M_p f(\theta) U^{m_2} \tag{4}$$

Where M_p is the weight of particles per unit volume of solution, $f(\theta)$ is a dependent function that determines the rate of erosion according to the angle of impact, and m_1 and m_2 are constants dependent on erosion-corrosion systems. Some researchers also found a power law between erosion-corrosion speed and flow speed and found that the power value of this law is between 1 and 4 [2].

Many metals and alloys have oxide films on their surface, such as Al_2O_3, Cr_2O_3, TiO_2, Fe_2O_3, Fe_3O_4, and NiO, and erosion leads to the removal of these films. If the flow rate is not high, the oxide films can reappear every time after being peeled off from the metal surface. If the flow rate exceeds some critical limits, there will not be enough time for the protective films to reestablish and remove the ions that will occur from the metal grid without the outer protective layer. In other words, the speed of removing the film is faster than the speed of its reproduction. Therefore, it may be easy to find different critical limits for water flow velocity for different metals and alloys on paper, but these limits must be used very carefully because the slightest change in the flow pattern, temperature, and chemical content of liquids and Alloys can lead to a change in the intensity of the critical flow rate [48].

In general, it can be said that increasing the flow rate, depending on its effect on the corrosion mechanism, may increase or decrease the corrosion rate. It can increase the rate of steel corrosion by bringing more oxygen, carbon dioxide, or hydrogen sulfide to the surface, and in the presence of corrosion inhibitors, increasing the flow rate can reduce the corrosion rate by bringing the inhibitors to the metal surface at a higher speed. It has been shown that to protect steel in drinking water at high flow rates, a small amount of sodium nitrite (inhibitor) is needed. Similar mechanisms have been proposed for other types of inhibitors.

The studies [40] of erosion-corrosion of aluminum and stainless steel alloys in fuming nitric acid have produced unexpected and interesting results, and with the increase of the flow rate, the corrosion of aluminum increased and the corrosion of stainless steel 347 decreased. The reason for this behavior was that the corrosion mechanisms in the two cases were different. **Figure 4** shows the increase in the corrosion rate of aluminum with an increase in the flow rate. In nitric acid that fumes aluminum, it forms aluminum nitrate and aluminum oxide. In a static state or at very low speeds, the corrosion rate is low or zero. At moderate velocities of 1–4 ft./sec, the nitrate shell is removed but not enough to remove the sticky oxide shell. Velocities above 4 ft./sec also destroy most of the oxide shell, and erosion-corrosion occurs at a faster rate.

Figure 5 shows the reduction of the corrosion rate of 347 stainless steel with an increasing flow rate. Under static conditions, this steel is auto-catalytically corroded

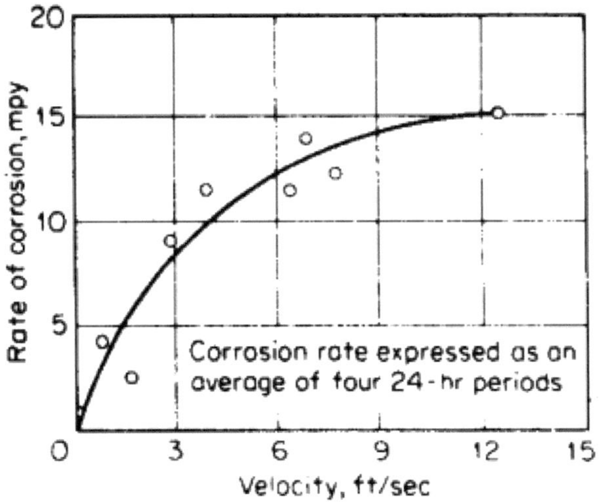

Figure 4.
Erosion corrosion of aluminum 3003 by fuming nitric acid at 108°F. Corrosion rate based on the average of four periods of 24 hours [40].

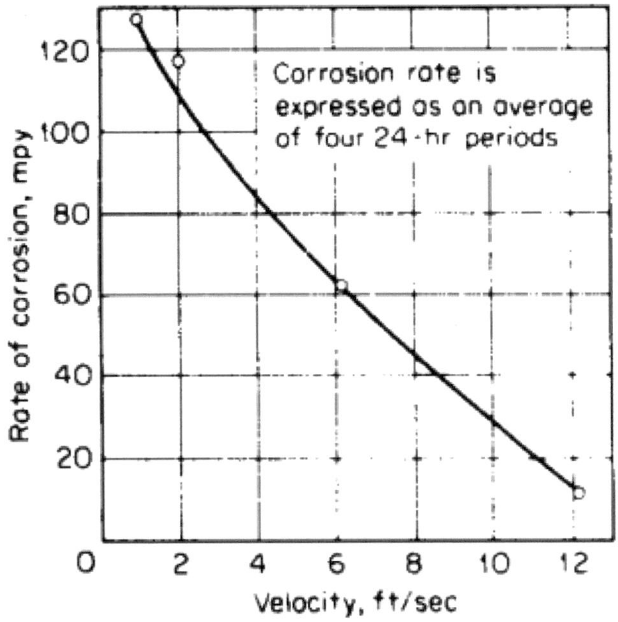

Figure 5.
Erosion corrosion of 347 stainless steel by white fuming nitric acid at 108°F. Corrosion rate based on the average of four periods of 24 hours [40].

by nitric acid because the cathodic reaction forms nitro acid. Increasing the flow rate causes nitro acid to leave the environment and reduces corrosion. It means at higher rates, the corrosion rate reduces by preventing sludge deposition, which would cause crevice corrosion in the absence of high rates.

Most stainless steels are prone to pitting and crevice corrosion in seawater and other chlorides, but some of these steels have been used successfully in seawater at high flow rates. This mode prevents the formation of sediments and prevents the initiation of cavities. Rate changes can also produce strange galvanic effects. In slow-moving seawater, the corrosion of steel does not change appreciably when in contact with stainless steel, copper, nickel, or titanium. At high flow rates, the corrosion of steel in contact with stainless steel and titanium is much less than when it is in contact with copper and nickel. This property is attributed to the more effective cathodic polarization of stainless steel and titanium at high rates [40].

2.3.2.2 Particle concentration

In low particle concentrations, an increase in the erosion rate has been observed linearly with increasing particle concentration. For high-concentration slurry, the erosion rate will also increase with increasing particle concentration, but it will be exponential and progressive rather than linear. Hutchings [28] found that this exponential movement in higher concentrations of particles is due to the interference of particles among each other. In lower particle concentrations, because the particles are almost independent of each other, the erosion rate increases linearly. He investigated the material loss behavior of mild steel in a slurry of silica particles and found that the lower the particle volume concentration from 12%, the material loss increases linearly.

When corrosion is present, the situation becomes more complicated. Zhu and his colleagues [18, 26] investigated the effect of particle concentration on corrosion current density in both active and inactive system states. For active systems [26], he studied the concentration effect on AISI 1045 low carbon steel at different rotation speeds with weight concentrations of 0, 20, and 35% and found that the current density significantly does not change with different concentrations of solid particles. For passive systems [18], the corrosion current density is highly dependent on the particle concentration, so the current density increases with the increase of the particle concentration.

2.3.2.3 Particle properties

The erosion-corrosion process depends on the properties of the particles. The rate of erosion is strongly affected by the hardness, size, and shape of the particles. Hutchings [42] found that particles that are 1.2–1.5 times harder than the surface of the target material produce a higher erosion rate than other particles. Oka and his colleagues [49] investigated the dependence of the particle size on the erosion rate of steel when its surface was hit by quartz particles and found that the erosion rate tends to increase up to a critical value, and after that, the erosion rate decreases to a relatively stable state.

2.3.2.4 Angle of impact

Finnie [50] investigated the erosion behavior of aluminum and its oxide at contact angles from nearly 0 to 90 degrees and found that for aluminum, as a soft material, the maximum erosion rate occurs between 15 and 30 degrees, while for aluminum oxide, as a hard material, the highest erosion rate occurs at 90 degrees. He explained that this angle change is due to the change in the erosion mechanism from soft to hard material. Erosive weight loss of soft materials is mainly done by grooving and cutting.

The reduction in the weight of hard materials is due to failure mechanisms because at 90 degrees, hard materials have the lowest resistance to failure; hence, we have the highest erosion rate.

Collision angles will also affect the corrosion process. Burstein and his colleagues [51] investigated the effect of collision angles on the increase of corrosion current, which is due to the breakdown or removal of the oxide layer on the surface.

2.3.3 Electrochemical parameters

2.3.3.1 The PH of the slurry

The PH of the slurry can greatly change the erosive corrosion behavior. As hydrogen ions are highly reactive, they can rapidly react with metal ions, resulting in severe surface erosion and reduced erosive resistance in anodic solutions. Guo and his colleagues [52] investigated the erosion-corrosion behavior of low carbon steel in abrasive slurries with pH 4, 7, and 10, and they found that the erosion rate was the highest in the solution with pH = 4 and the lowest in the solution with pH = 7 and at PH = 10, and our speed was between these two. Zhou et al. [53] also studied the effect of pH on the erosive corrosion characteristics of ductile cast iron under an open-circuit potential and obtained similar results. They explained that slurry with lower pH results in a higher corrosion current density under an open-circuit potential; hence, the erosion resistance will be lower.

2.3.3.2 Corrosion current density

Usually, the rate of erosion, in the case of corrosion, increases with the corrosion current density. Lu and his colleagues [45] investigated the effect of corrosion current density on the erosion-corrosion rate and found that the erosion rate increases linearly with the logarithm of the anodic current density. He also proposed an equation to calculate the theory of increased erosion rate due to corrosion with a given anodic current density.

$$\frac{e_c}{e_\circ} = z \log\left(\frac{i_a}{i_{th}}\right) \tag{5}$$

Where i_a and i_{th} are the actual anode current density and the threshold density that causes mechanical destruction of the surface, respectively. Z is a constant that varies between erosion-corrosion systems.

2.4 Erosion corrosion mechanism

If liquids and gases containing solid particles or gases containing liquid droplets flow on the surface of the metal, the abrasive agent causes mechanical wear of the metal. As a result, the metal decomposes in the form of ions or the formation of corrosion products.

The erosion-corrosion mechanism depends on the following components:

1. Flow speed and its characteristics: flow geometry (turbulent or smooth), the presence of obstacles in front of the flow, the angle of the flow to the metal surface, etc.

2. Environmental conditions (mechanical and physical characteristics): PH, presence of water droplets in steam flow, acidic gases, two-phase water or water-steam environment, etc.

3. The nature of metals and alloys: chemical composition, hardness, metallurgical factors (the presence of various phases such as ferrite, bainite, martensite, and austenite in steels), the type and structure of passive films on the surface of metals.

In general, there are two main reasons for the occurrence of erosion-corrosion. The first reason is "Erosion" due to the impact of certain materials or the impact of certain drops on the surface of metals. Since ancient times, humans have learned to cut "weak" materials such as wood and leather using "hard" materials such as stone and metal. The same phenomenon occurs on the surface of metals. If two solid materials that have hardness are different (metals and corroded particles) and come into contact with each other and move in opposite directions, the material with higher hardness will scratch the other surface.

The kinetic energy of certain materials (solid particles) and liquid droplets that move at high speed carries the necessary energy to cut or break the outer layer of metals. "Erosion" is a mechanical action of "wearing" metal. Corrosion is a chemical action of metal dissolution. Increasing the current speed increases the speed of transferring aggressive components to the metal surface and corrosion products in the opposite direction from the metal surface to the environment; as a result, the penetration speed of the components participating in the cathode and anode of corrosion reactions increases. The use of electrochemical methods for monitoring and controlling erosion-corrosion shows an aspect of the electrochemical mechanism in this complex phenomenon.

The second reason for the occurrence of erosive corrosion is cavitation, which is the formation and collapse of bubbles in the liquid (the first type of cavitation), or the condensation of vapor molecules (the second type of cavitation) on the metal surface during the flow. In the first type of cavitation phenomenon, the turbulent flow of liquids (strong flow under turbulence) causes a change in the pressure in the liquid flow near the metal surface. This condition can occur on the surface of a ship propeller or in centrifugal pumps. The collapse of vapor molecules forms local stresses on the surface of the metal due to the released shock wave. These stresses have higher energy and forces than the chemical electrostatic forces between atoms in the metal lattice. The surface of the metal is not uniform and forms surface pits such as cavities. These holes are called the pitting phenomenon. Chemical factors (if corrosive chemicals are present in the liquid or the vapor stream) can accelerate pitting [54].

Conflict of interest

"The author declares no conflict of interest."

Author details

Sajjad Akramian Zadeh
University of Razi, Kermanshah, Iran

*Address all correspondence to: sajjadakramian@gmail.com

References

[1] Babu R et al. Erosion corrosion of HSLA steel in a coal slurry. Corrosion Prevention & Control. 1996;**43**(5):131-134

[2] Wang K. The Effect of Solution Composition on Erosion-corrosion Process and Correlation Between Repassivation Kinetics and Corrosion Rate in Flowing Slurry. Alberta, Canada: University of Alberta; 2008

[3] Bradford SA, Bringas JE. Corrosion Control. Vol. 115. Berlin, Germany: Springer; 1993

[4] Jones DA. Principles and Prevention of Corrosion. Upper Saddle River, USA: Macmillan; 1992

[5] Weber J. Flow induced corrosion: 25 years of industrial research. British Corrosion Journal. 1992;**27**(3):193-199

[6] Heitz E. Chemo-mechanical effects of flow on corrosion. Corrosion. 1991; **47**(2):135-145

[7] Stachowiak G, Batchelor A. Engineering Tribology. Burlington, Mass, USA: Elsevier, Butterworth-Heinemann; 2005

[8] Bardal E. Corrosion and Protection. Springer Science & Business Media; 2007

[9] Matsumura M et al. The role of passivating film in preventing slurry erosion-corrosion of austenitic stainless steel. ISIJ International. 1991;**31**(2): 168-176

[10] Li Y, Burstein G, Hutchings I. The influence of corrosion on the erosion of aluminium by aqueous silica slurries. Wear. 1995;**186**:515-522

[11] Yue Z, Zhou P, Shi J. Some factors influencing corrosion–Erosion performance of materials. Wear of Materials. 1987;**1987**(2):763-768

[12] Zheng Y et al. The synergistic effect between erosion and corrosion in acidic slurry medium. Wear. 1995;**186**: 555-561

[13] Ashby M, Lim S. Wear-mechanism maps. Scripta Metallurgica et Materialia. 1990;**24**(5):805-810

[14] Li W, Li D. Influence of surface morphology on corrosion and electronic behavior. Acta Materialia. 2006;**54**(2): 445-452

[15] Li W, Li D. On the correlation between surface roughness and work function in copper. The Journal of Chemical Physics. 2005;**122**(6):064708

[16] Sasaki K, Burstein G. The generation of surface roughness during slurry erosion-corrosion and its effect on the pitting potential. Corrosion Science. 1996;**38**(12):2111-2120

[17] Wang X, Li D. Application of an electrochemical scratch technique to evaluate contributions of mechanical and electrochemical attacks to corrosive wear of materials. Wear. 2005;**259**(7): 1490-1496

[18] Guo H, Lu B, Luo J. Study on passivation and erosion-enhanced corrosion resistance by Mott-Schottky analysis. Electrochimica Acta. 2006; **52**(3):1108-1116

[19] Zhao X et al. Effects of erosion on corrosion of type 430 and 316 stainless steels in aqueous environments. British Corrosion Journal. 2002;**37**(1):63-68

[20] Adler T, Walters R. Repassivation of 304 stainless steel investigated with a

single scratch test. Corrosion. 1993; **49**(5):399-408

[21] Burstein G, Sasaki K. Detecting electrochemical transients generated by erosion–corrosion. Electrochimica Acta. 2001;**46**(24):3675-3683

[22] Yin S, Li D. Effects of prior cold work on corrosion and corrosive wear of copper in HNO3 and NaCl solutions. Materials Science and Engineering: A. 2005;**394**(1–2):266-276

[23] Li J et al. Depassivation and repassivation of AISI321 stainless steel surface during solid particle impact in 10% H2SO4 solution. Wear. 1995;**186**: 562-567

[24] Mayuzumi M, Ohta J, Arai T. Effects of cold work, sensitization treatment, and the combination on corrosion behavior of stainless steels in nitric acid. Corrosion. 1998;**54**(4):271-280

[25] Stack M, Jana B. Modelling particulate erosion–corrosion in aqueous slurries: Some views on the construction of erosion–corrosion maps for a range of pure metals. Wear. 2004;**256**(9–10): 986-1004

[26] Guo H, Lu B, Luo J. Interaction of mechanical and electrochemical factors in erosion–corrosion of carbon steel. Electrochimica Acta. 2005;**51**(2): 315-323

[27] Postlethwaite J. Effect of chromate inhibitor on the mechanical and electrochemical components of erosion-corrosion in aqueous slurries of sand. Corrosion. 1981;**37**(1):1-5

[28] Hutchings IM. Monograph on the Erosion of Materials by Liquid Flow. MTI Publication, Material Technology Institute of the Chemical Process Industries; 1986

[29] Saraswathi Y, Das S, Mondal D. Erosion-corrosion behavior of SiC particle-reinforced Al-Si alloy in NaOH slurry. Metallurgical and Materials Transactions A. 2005;**36**(8):2259-2262

[30] Reyes M, Neville A. Mechanisms of erosion-corrosion on a cobalt-base alloy and stainless-steel UNS S17400 in aggressive slurries. Journal of Materials Engineering and Performance. 2001; **10**(6):723-730

[31] Guo H, Lu B, Luo J. Response of surface mechanical properties to electrochemical dissolution determined by in situ nanoindentation technique. Electrochemistry Communications. 2006;**8**(7):1092-1098

[32] Jones DA, Jankowski AF, Davidson GA. Room-temperature diffusion in Cu/Ag thin-film couples caused by anodic dissolution. Metallurgical and Materials Transactions A. 1997;**28**(13):843-850

[33] Caceres C, Selling B. Casting defects and the tensile properties of an Al Si Mg alloy. Materials Science and Engineering: A. 1996;**220**(1):109-116

[34] Gutman EM. Mechanochemistry of Materials. Cambrige, United Kingdom: Cambridge Int Science Publishing; 1998

[35] Gu B et al. The effect of anodic polarization on the ambient creep of brass. Corrosion Science. 1994;**36**(8): 1437-1445

[36] Wang B, Geng G, Levy AV. Effect of microstructure on the erosion-corrosion of steels. Wear. 1991;**151**(2):351-364

[37] McCabe LP, Sargent GA, Conrad H. Effect of microstructure on the erosion of steel by solid particles. Wear. 1985; **105**(3):257-277

[38] Lindsley B, Marder A. Solid particle erosion of an Fe-Fe3C metal matrix composite. Metallurgical and Materials Transactions A. 1998;**29**(3): 1071-1079

[39] Bregliozzi G et al. Cavitation erosion resistance of a high nitrogen austenitic stainless steel as a function of its grain size. Journal of Materials Science Letters. 2003;**22**(13):981-983

[40] Fontana MG. Corrosion Engineering. McGraw Hill, India: Tata McGraw-Hill Education; 2005

[41] Metwally W, Samy M. Evaluation of abrasive-wear and erosion-corrosion resistance of high-Cr cast-steel. Steel Research. 1994;**65**(10):455-458

[42] Hutchings I. A model for the erosion of metals by spherical particles at normal incidence. Wear. 1981;**70**(3): 269-281

[43] Oka YI, Matsumura M, Kawabata T. Relationship between surface hardness and erosion damage caused by solid particle impact. Wear. 1993;**162**:688-695

[44] Finnie I. Some observations on the erosion of ductile metals. Wear. 1972; **19**(1):81-90

[45] Lu B, Luo J. Synergism of electrochemical and mechanical factors in erosion-corrosion. The Journal of Physical Chemistry B. 2006;**110**(9): 4217-4231

[46] Meng H, Ludema K. Wear models and predictive equations: Their form and content. Wear. 1995;**181**:443-457

[47] Postlethwaite J, Dobbin MH, Bergevin K. The mechanism of Erosion-corrosion in slurry pipelines. In: Materials Science Forum. Baech, Switzerland: Trans Tech Publ; 1986

[48] Groysman A. Corrosion for Everybody. Berlin, Germany: Springer Science & Business Media; 2009

[49] Oka Y, Matsumura M. Erosive wear testing apparatus–Simulation of erosion caused by slurry of low-impingement velocity. Wear of Materials. 1983;**1983**: 360-366

[50] Finnie I. Erosion of metals by solid particles. Journal of Materials. 1967;**2**:682

[51] Burstein G, Sasaki K. Effect of impact angle on the slurry erosion–corrosion of 304L stainless steel. Wear. 2000;**240**(1–2):80-94

[52] Guo H, Lu B, Luo J. Non-Faraday material loss in flowing corrosive solution. Electrochimica Acta. 2006; **51**(25):5341-5348

[53] Zhou Y, Lu Z, Zhan M. An investigation of the erosion–corrosion characteristics of ductile cast iron. Materials & Design. 2007;**28**(1):260-265

[54] Groysman A, Groysman O. Corrosion for Everybody. Berlin, Germany: Springer; 2009

Chapter 8

Corrosion Protection and Modern Infrastructure

Sameer Dohare

Abstract

Currently, modern infrastructures utilize different materials such as metals, alloys, glass, plastic, wood, ceramic, silicate brick, and natural stones. The material's diverse composition, structure, chemical, physical, and mechanical properties with ease in utilization make metal utilization a priority for architects and civil contractors. The principal property that determines the quality and durability of infrastructure is the corrosion resistance and weathering resistance. The corrosion of metals can cause damage to concrete, building stones, wood, and other materials, leading to corrosive destruction. The corrosion protection of the structural steel used in reinforcements, load supports, and frames has become extremely important to meet the demand of modern infrastructure having 100 to 120 years of service life.

Keywords: corrosion, protection, modern infrastructure, equipment, structural steel, cable bridges, suspension cable, galvanized steel, polymeric coating

1. Introduction

The term corrosion originated from the Latin word corrodere, meaning gnawing to pieces. Corrosion is the process of material deterioration due to electrochemical interactions with the surrounding environment.

Since ancient times, metallic corrosion has been a problem in the utilization of metals. Nearly all the metals present in nature are in the metal compounds, such as oxides, silicates, and carbonates. Hence, after the extraction of metals, these metals tend to become naturally occurring metal compounds once in the environment.

One of the most known examples of corrosion is rusting of iron. The rusting of iron is the formation of iron oxides. The extraction of metal iron from its iron oxides present in the earth's crust is just a reverse process of the rusting of iron. The reaction of iron with oxygen to form iron oxide is a reversible reaction.

Considerable efforts had made to develop corrosion prevention measures such as metal doping and coatings of other metals, pigments, polymers, and organic materials. However, due to the limitations of these preventive measures, complete corrosion prevention is still unachievable. The main drawback of structure corrosion is that it leads to structure failure catastrophe. One of the well-known catastrophes is the sinking of the Titanic. The utilization of different types of iron in about 3 million rivets in the highly corrosive seashore environment leads to the corrosion of rivet joints. These weakened joints aided in the catastrophe.

As the demand for high-strength and lightweight structures increases, metals become one of the most favorable choices for the structures.

Currently, modern infrastructures utilize different materials, such as metals, alloys, glass, plastic, wood, ceramic, silicate brick, and natural stones, increasing the corrosion prevention challenge. The benefits of the metal's diverse composition, structure, and chemical, physical and mechanical properties ease metal utilization for modern infrastructure, making corrosion prevention a challenging task.

A combination of incompatible materials in the environment leading to the decline in corrosion resistance and weathering resistance of the infrastructure will reduce the life of the infrastructure. There is a need for the corrosion protection of structural steel, especially for the reinforcements, load supports, and frames that have become extremely important to meet the demand of modern infrastructure having 100–120 years of service life.

In the following sections, we will focus on the corrosion causes in steel reinforcement, steel cables, and structural steel with the advantages and disadvantages of different corrosion protection techniques.

2. Corrosion protection techniques

As discussed earlier, metal corrosion is an electrochemical reaction that requires water, oxygen, and ions like chloride ions, which already exist in the atmosphere. These atmospheric ions are abundant near the coastline as the air carries these ions from the available saline water (**Figure 1**).

Moreover, pollutants such as carbon dioxide (CO_2), carbon monoxide (CO), sulfur dioxide (SO_2), and nitrous oxide (NO_2) present in the environment also be significant in the corrosion process. **Figure 2** demonstrates the electrochemical reactions of corrosion of the iron metal.

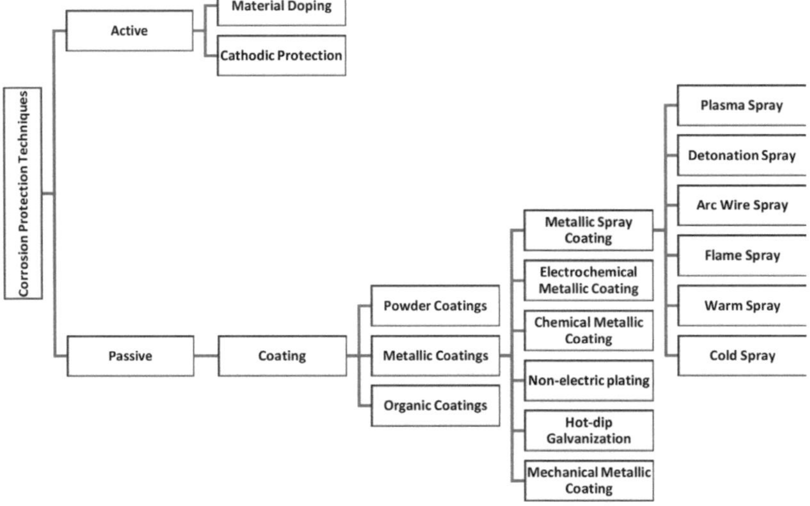

Figure 1.
Corrosion protection techniques.

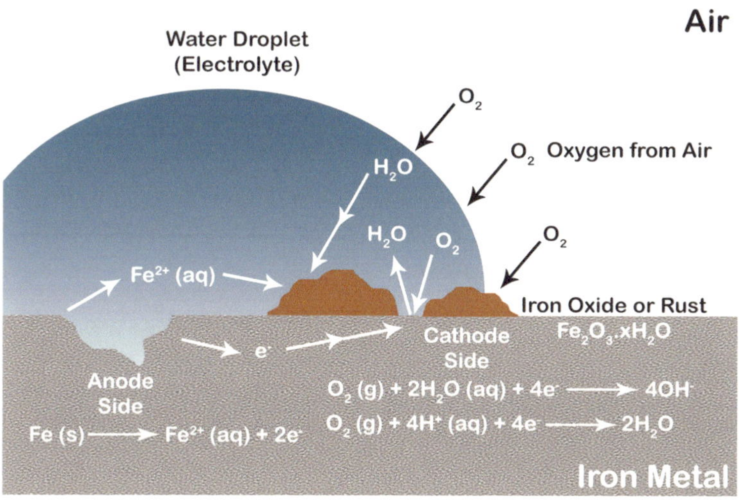

Figure 2.
Corrosion of iron metal.

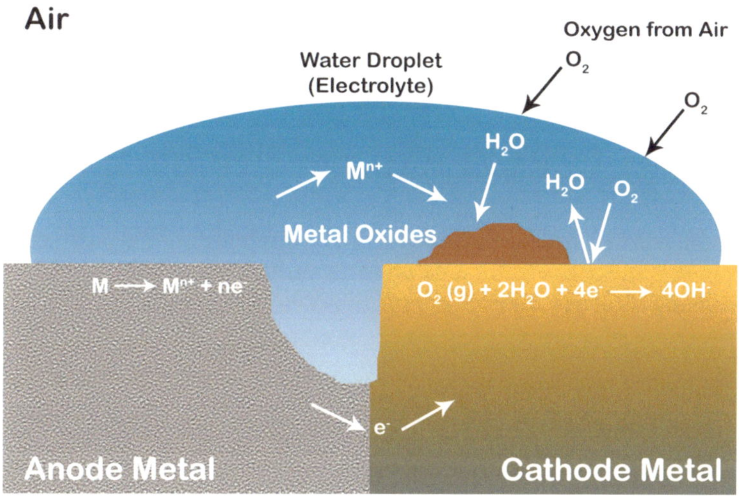

Figure 3.
Galvanic corrosion.

This multifaceted phenomenon of corrosion adversely affects and causes the deterioration of materials. Millions of dollars are infused throughout the infrastructure sector from corrosion prevention and control.

When two or more dissimilar materials having different potentials come in contact with each other in the conductive electrolyte, due to the generation of current, the more reactive metal will corrode in preference to the less reactive metal. Corrosion will occur at the point where the current leaves the metal surface. **Figure 3** shows the basics of galvanic corrosion.

Before going into detail about the causes of corrosion in steel reinforcement, steel cables, and structural steel, let us discuss the types of corrosion protection techniques.

2.1 Active corrosion protection techniques

Active corrosion protection techniques focus on halting or neutralizing corrosion electrochemical reactions. The active corrosion protection techniques inhibit corrosion on the material to be protected. It is the application of the reactive compound to disrupt the normal formation of anodes on the materials. **Figure 4** shows the basics of the active corrosion protection mechanism. The more reactive metal becomes the sacrificial anode to protect the less reactive metal, which acts as a cathode.

2.2 Passive corrosion protection techniques

Passive corrosion protection techniques focus on the isolation of the material from corrosion-causing elements to constraint corrosion. With passive protection, a protective coating, for example, may act as a barrier that prevents air and moisture from coming into contact with the underlying iron substrate. With these two elements out of the picture, corrosion cannot occur on the surface of the metal. **Figure 5** shows the basic mechanism of the passive corrosion protection technique.

In the active corrosion protection technique, the material remains exposed to corrosion-causing elements while various processes actively counteract the material corrosion. However, passive corrosion protection techniques involve the separation of material from the corrosion-causing elements.

2.3 Active corrosion protection techniques

As discussed earlier, one of the most adopted methods of active corrosion protection techniques is the use of cathodic corrosion protection. Cathodic corrosion protection involves a direct or indirect connection with a more reactive material to the

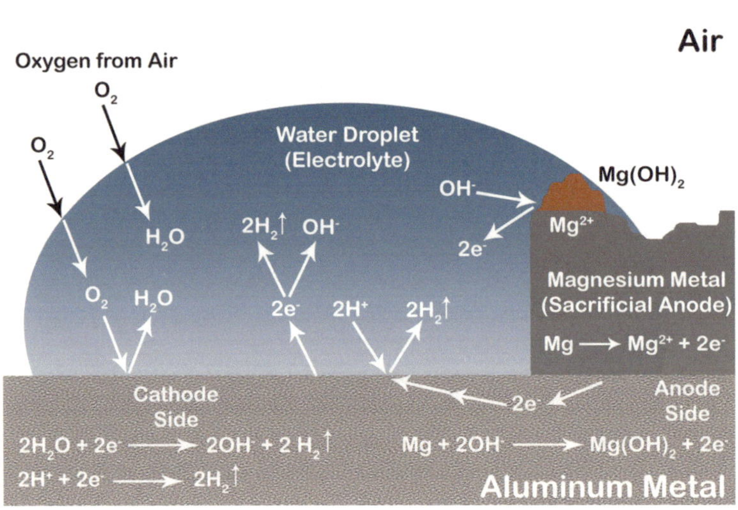

Figure 4.
Active corrosion protection.

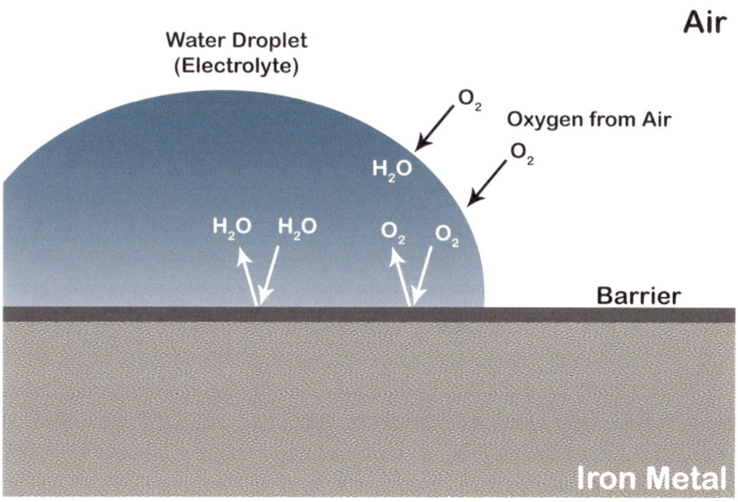

Figure 5.
Passive corrosion protection.

material to be protected. It is a simple method of diverting the corrosion to the sacrificial material while the other material remains protected.

For example, the addition of inorganic zinc inhibitive pigments such as zinc phosphate ($Zn_3(PO_4)_2$) in steel offers active anticorrosive protection to the steel substrate by hydrolyses in water to produce zinc ions (Zn_2^+) and phosphate ions (PO_4^{3-}). The zinc and phosphate ions act as cathodic and anodic inhibitors, respectively. The phosphate ions phosphating the steel and rendering it passive is another advantage of using zinc phosphate. **Figure 6** shows the mechanism of the zinc phosphate corrosion inhibitor.

The following are the principal active corrosion protection techniques.

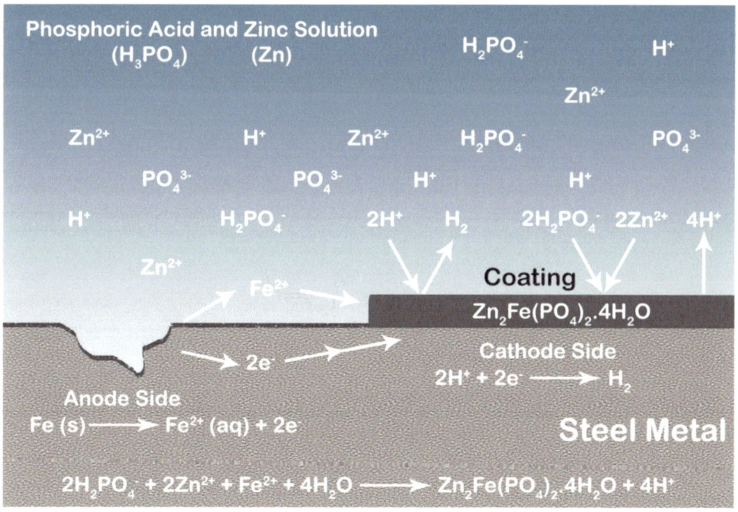

Figure 6.
Corrosion inhibitor.

2.3.1 Material doping

Material doping, also known as alloying, is a method in which one or more elements or compounds are doped into the material to change material properties, like increased corrosion resistivity. Alloying is the most effective method to control corrosion. Humankind has already revolutionized the world various times by developing alloys such as bronze, steel, brass, alnico, nichrome, cast iron, and carbon steel. Stainless steel is a mixture of iron, chromium, carbon, nickel, molybdenum, titanium, niobium, manganese, and more. The stainless steel is stainless material, not fully corrosion-resistant. However, due to the added strength and resistance to corrosion of stainless steel, stainless steel is preferred over other materials for construction in the infrastructure segment.

For example, nickel has good corrosion-resistant properties, and chromium has good oxidation-resistant properties. When nickel and chromium doped into the material, the resultant alloy gives the best resistance in highly oxidized and reduced chemical environments. **Figure 7** shows the principal mechanism of chromium and nickel doping in the iron to form iron alloy or steel. Chromium and nickel having inherent corrosion-resistive properties develop the protective oxide layer, and change the crystalline structure of iron from ferritic (Body Centered Cubic Crystal) structure to austenitic (Face Centered Cubic Crystal) structure, respectively.

Different alloys provide different resistance to different environments. However, despite the effectiveness of alloys, the doping process makes them very expensive. Sometimes so expensive that the replacement cost of the highly corroded complete structure becomes economical.

2.3.2 Cathodic protection

Cathodic protection is one of the most effective methods used for corrosion control. Cathodic protection protects the material by converting the active sites of the material to passive sites by providing electrons from galvanic anodes attached to or near the material. Generally, used materials for galvanic anodes are aluminum, magnesium, or

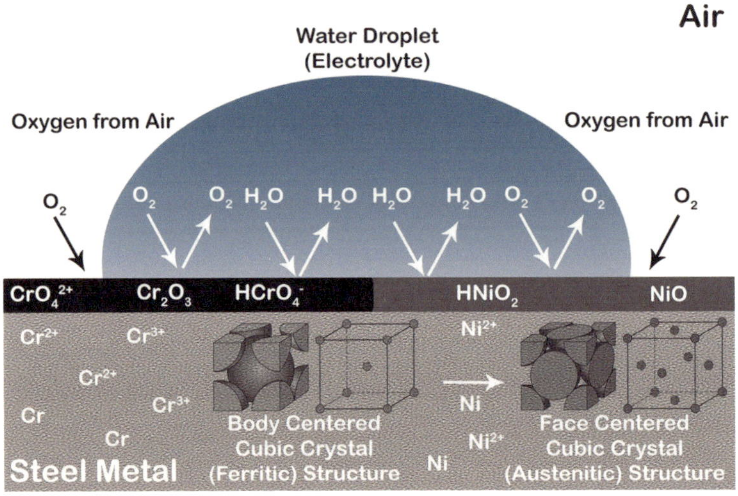

Figure 7.
Metal doping corrosion protection.

zinc. Zinc is the most widely used metal for the protection of steel, as zinc metal in direct contact with steel offers protection through the preferential oxidation of zinc metal. The rate of corrosion of zinc is also slow compared with steel. **Figure 8** demonstrates the basic electrochemical corrosion reactions of zinc metal anode to protect the iron metal. However, in the presence of ions such as chlorides in coastal regions, this reaction rate gets accelerated, limiting the zinc protection use in the coastal area.

Cathodic protection is highly effective, but the high anode consumption requires frequent checks and replacements, increasing the cost of maintenance. Further, an anode increases the overall weight structure and is ineffective in high-resistivity environments, constraining the utilization of cathodic protection. **Figure 9** demonstrates the basic electrochemical corrosion reactions of sacrificial zinc metal anode to protect the reinforcement steel bars of the pile foundation.

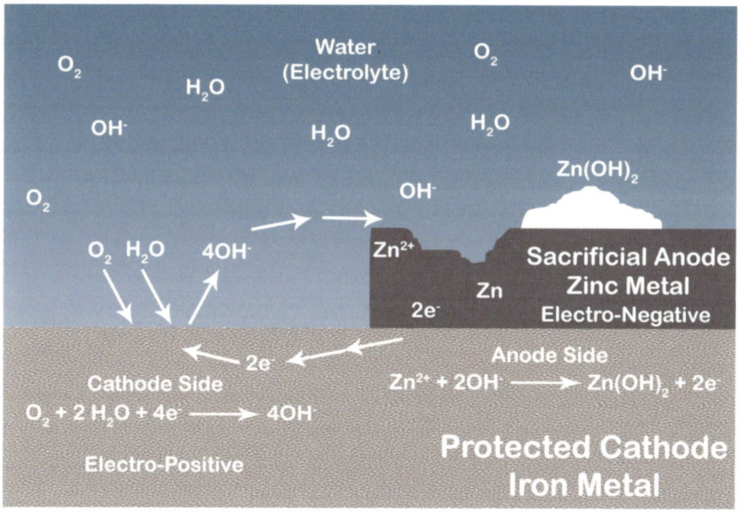

Figure 8.
Cathodic protection of iron metal.

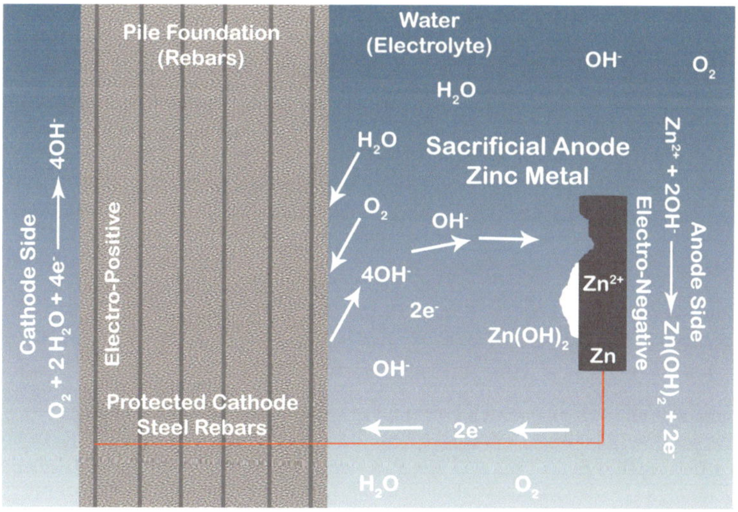

Figure 9.
Cathodic protection of reinforcement steel bars.

Cathodic protection is adopted globally to protect offshore production platforms, pipelines, water storage tanks, water treatment plants, boat hulls, ships, piers, reinforcement bars in concrete structures, and more.

2.4 Passive corrosion protection techniques

In passive corrosion protection techniques, the corrosion damage is prevented by mechanically isolating the materials, using protective layers, films, or coatings, from the corrosion-causing elements. Passive corrosion protection techniques neither change the corrosion resistivity of the material nor the corrosivity of the corrosion-causing elements. The main drawback of passive corrosion protection techniques is that at any point if the protective layer, film, or coating is destroyed or damaged, the corrosion of the material will occur. Passive corrosion protection techniques are used to protect the material at the place of use for relatively mild environmental conditions. Harsh environmental conditions generate stresses and reduce the effectiveness of the protective layer, film, or coating.

For example, metal oiling is one of the best and most conventional methods used for corrosion protection. The protective layer of oil does not allow water or hydrophilic electrolytes to complete the electrochemical reaction of the corrosion. Further, the penetration of oil into holes, cavities, and difficultly accessible areas makes the corrosion protective layer more efficient. However, oiling is avoided in water-submerged applications and high hygiene or safe working environment area. The following are the principal passive corrosion protection techniques.

2.4.1 Coating

This passive corrosion protection technique is based on providing a barrier coating to the material to prevent exposure to corrosion-causing elements, which are oxygen, water, and ions. **Figure 10** shows the basic composition of the paints. Painting is one

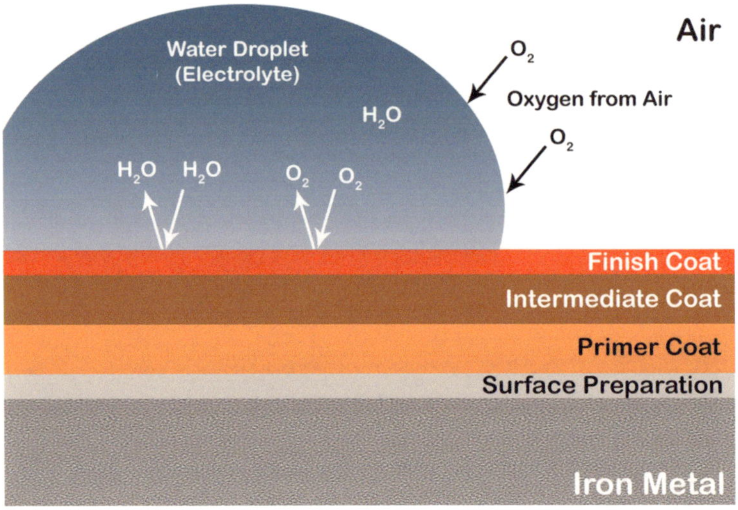

Figure 10.
Paint coatings.

of the easiest and cheapest ways to prevent corrosion. Paint acts as a barrier between material surface and corrosion-causing elements. The combination of different paint layers acts as a different corrosion protection function. The primer coat acts as an inhibitor, the intermediate layer provides strength, and the outer layer protects from the environment.

Based on the severity of the environment, various coatings can be applied. Powder coating, metallic coating, and organic coating are the principal coating types.

2.4.1.1 Powder coatings

The powder coating technique is a process of electrothermal fusion of a powder on the clean surface of the material to be protected. The dry powder is static electrically charged and deposited on the oppositely charged or grounded material forming a smooth and continuous film. This film along with the material is heated, and a protective layer of powder is fused with the material to protect from corrosion. This technique can provide coating thicknesses in the range of 25 to 125 micrometers. Generally, powders of acrylic, vinyl, epoxy, nylon, polyester, and urethane are used for coating. **Figure 11** shows the general powder coating process applied in the industries.

Compared with conventional liquid paint where paint is delivered through evaporation of the solvent, the powder coating is applied electrostatically and then cured under heat or with ultraviolet light, this creates a hard finish layer with durability, to withstand damage and last longer.

2.4.1.2 Metallic coatings

The metallic coating is preferable where the pores-free or damage-free coat of more noble material can be applied on the material to be protected from corrosion. These noble materials can be a metal or alloys. The metallic coating is applied using a sprayer, electrochemically, chemically, or mechanically.

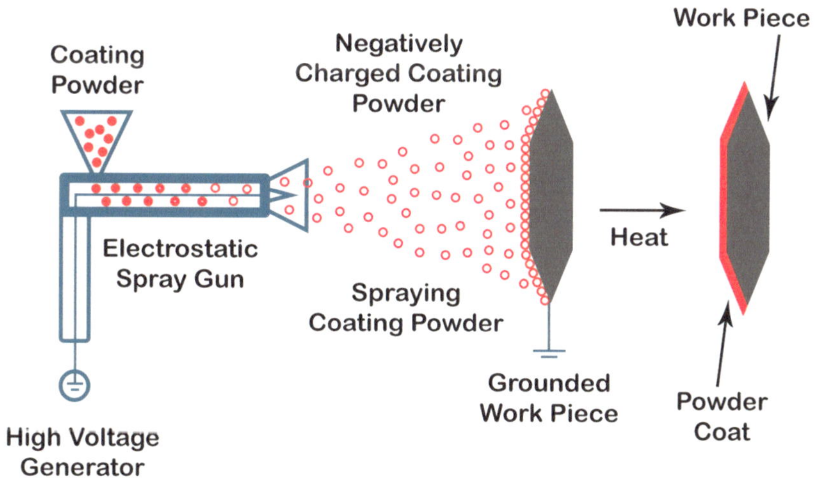

Figure 11.
Powder coating process.

2.4.1.2.1 Metallic spray coating

The metallic spray coating technique involves coating material in a molten or semi-molten state. The following are various metallic spray coating processes. **Figure 12** shows the general metallic spray coating process.

Plasma spray: The plasma spray coating process utilizes the plasma jet to melt the metallic powder coating material, which then sprays onto the material to be protected.

Detonation spray: The detonation spray coating process utilizes a very-high shockwave to coat molten or partially molten coating materials onto the surface of the material to be protected.

Arc wire spray: The arc wire spray coating process utilizes an electric arc to melt the metallic powder of coating material, which then pneumatically spray onto the surface of the material to be protected.

Flame spray: The flame spray coating process utilizes the flame to melt the metallic powder and compressed air to atomize and propel the coating material onto the surface of the material to be protected.

Warm spray: The warm spray coating process involves the deposition of heated metallic powder at supersonic speed onto the surface of the material to be protected.

Cold spray: The cold spray coating process utilizes a very high speed of carrier gas to generate high-impact forces on the metallic powder. These high-impact forces create a protective letter.

Metallic spray coating is used to coat material to protect against extremes of temperature, corrosion, erosion, and general wear and tear. Tungsten carbides, ceramics, nickel-chrome carbides, aluminum, steels, and plastics are some of the materials used to apply them as coating materials.

2.4.1.3 Electrochemical metallic coating

Electrochemical metallic coating, also known as electrocoating, is the process in which electrically charged particles are deposited on the material surface to form a

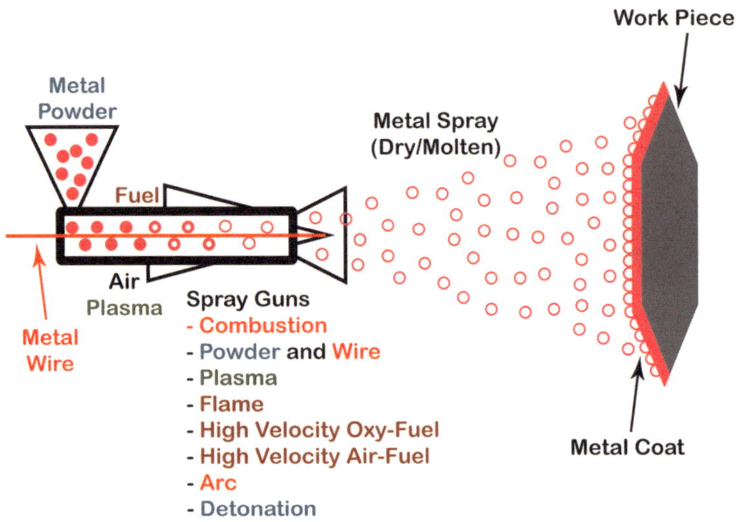

Figure 12.
Metallic spray coating process.

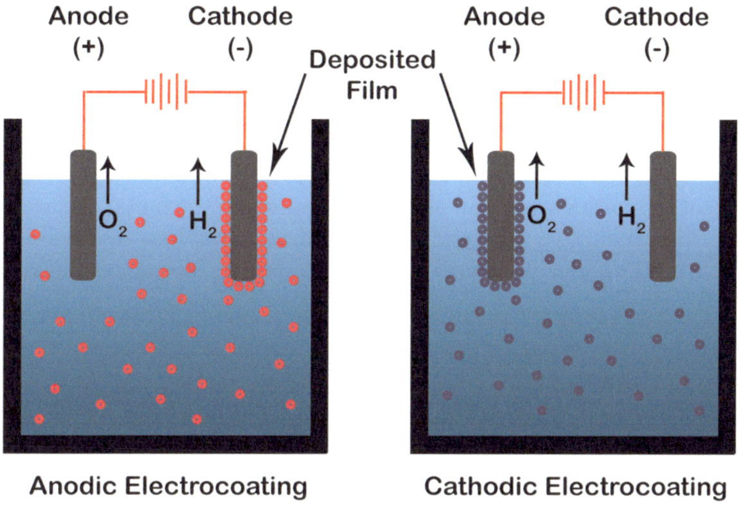

Figure 13.
Electrochemical metallic coating.

protective coating on the material to be protected. **Figure 13** shows the two types of generally used anodic electro-coating and cathodic electro-coating processes in the industries.

Generally, the metal ions deposited on the material are cadmium, chromium, nickel, and zinc. Electroplating provides very high control over protective coating. By controlling temperature, current, voltage, metal ion concentration, and solution of the coating tank, in which the material to be protected will be immersed, up to 1 μm of protective coating is possible.

2.4.1.4 Chemical metallic coating

Chemical metallic coating is a coating technique where the protective coat is chemically bounded with the material to be protected. There are two principal techniques for chemical coat materials. The nonelectric plating and hot-dip galvanization (**Figure 14**).

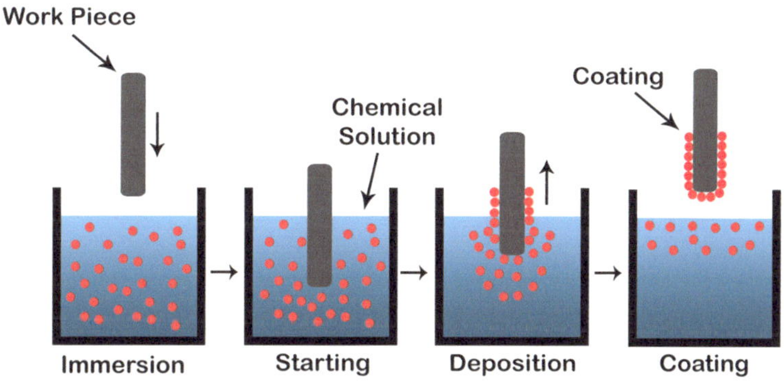

Figure 14.
Chemical metallic coating.

2.4.1.5 Nonelectric plating

Nonelectric or electroless plating is the process where the metal is deposited on the material *via* a chemical reaction in the presence of a catalyst. The most common nonelectric plating is electroless nickel plating. Electroless nickel plating is generally achieved by depositing the nickel ions with sodium hypophosphite as the reducing agent to oxidize and form a negative charge on the metal surface for the deposition of nickel ions. In this process, the sodium hypophosphite will release the hydrogen as a hydride ion. The overall reaction of electroless nickel plating is as follows.

$$\mathrm{NiLx}^{(2+)} + 4\mathrm{H_2PO_2^{(-)}} + \mathrm{H_2O} \rightarrow \mathrm{Ni} + \mathrm{P} + 3\mathrm{H_2PO_3^{(-)}} + \mathrm{H}^{(+)} + \frac{3}{2}\mathrm{H_2}\uparrow + \mathrm{xL} \quad (1)$$

As electroless plating chemically generates the charge on the surface of workpiece, it is highly recommended for irregularly shaped objects which are difficult to plate evenly with electroplating. Moreover, nonelectric plating is also used as a pretreatment unit to deposit a conductive surface on nonconductive surfaces, such as polymers, to electroplate the nonconductive materials.

2.4.1.6 Hot-dip galvanization

The hot-dip galvanization process has been around for more than 250 years. The hot-dip galvanization process is used for the corrosion protection of artistic sculptures. The hot-dip galvanization corrosion prevention method involves dipping the material into molten metal. The material reacts with the metal to create a tightly bonded coating. This chemically bonded coating acts as corrosion protection.

The generally used molten metal is zinc. When the steel is immersed in the molten bath of zinc, the zinc adheres to the steel. When the adhered zinc comes in contact with the air, it immediately reacts with the oxygen present in the air and forms a very strong zinc oxide layer, preventing corrosion. The zinc and steel form a metallurgical bond. Hence, the applied coating will not flake off. The hot or cold rolled coils are supplied with the metallic coating applied by either electroplating or hot dipping. The generally applied coating includes zinc, aluminum, tin, and lead.

2.4.1.7 Mechanical metallic coating

The mechanical metallic coating is the process of cold welding the fine metal powder on the material to be protected by tumbling the material with metal powder and a media in an aqueous solution. The mechanical metallic coating is generally used to apply zinc or cadmium to small parts as fasteners (**Figure 15**).

2.4.1.8 Organic coating

The organic coating technique is a process of coating material by utilizing carbon-rich compounds to get a monolithic or multilayer protection coat. Such compounds are generally obtained from vegetables or animals. The organic coating thickness of it

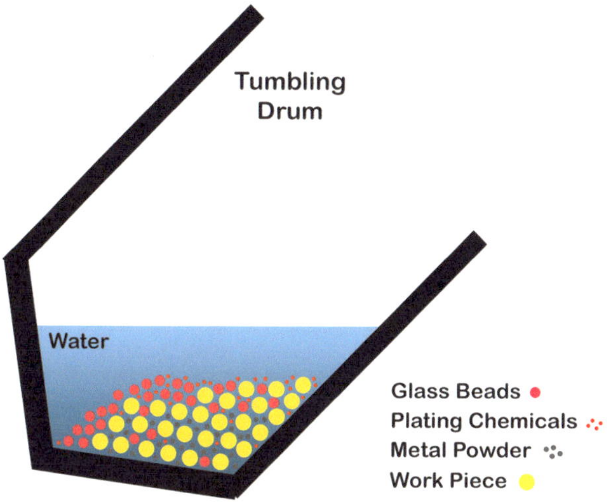

Figure 15.
Mechanical metallic coating.

depends on time and temperature. Initially, the organic coating deposition is rapid but slows down as the coating begins to build or get mature, that is, the rate of coating declines with time and thickness. The coating thickness from 15 to 25 μm is achievable in the organic coating technique.

3. Corrosion causes in steel reinforcement

Concrete is a construction material composed of fine and coarse aggregate bonded with cement and water, which hardened over time. Concrete has very high compressive strength, but its low tensile strength limits the utilization of concrete in the infrastructure sector. In the fifteenth century, the introduction of steel reinforcement changed the infrastructure sector forever.

Steel reinforcement, also known as rebar, is the steel products such as mesh, wire, or bars used in the concrete to increase the tensile strength by holding the concrete in tension. As the steel reinforcement acts as a tensile device and significantly increases the tensile strength of concrete, the utilization of steel-reinforced concrete in the infrastructure sector also increased. However, this reinforcement inside the reinforced concrete structure is susceptible to corrosion damage.

When the iron gets oxidized to iron oxide, this iron oxide forms a layer around the bar, causing the expansion of the rebar. This expansion set up internal stress in the concrete, leading to cracks in the concrete. This corrosion seriously damages the structure and may lead to total structural collapse.

To understand the causes, it is recommended to understand the concrete cement process. Hydration or curing is the process of hardening concrete with water. **Figure 16** shows the basics of the hydration process. During the hydration process, the compounds in the cement form chemical bonds with water molecules and become hydrates or hydration products. The following chemical reactions occur during the hydration of cement.

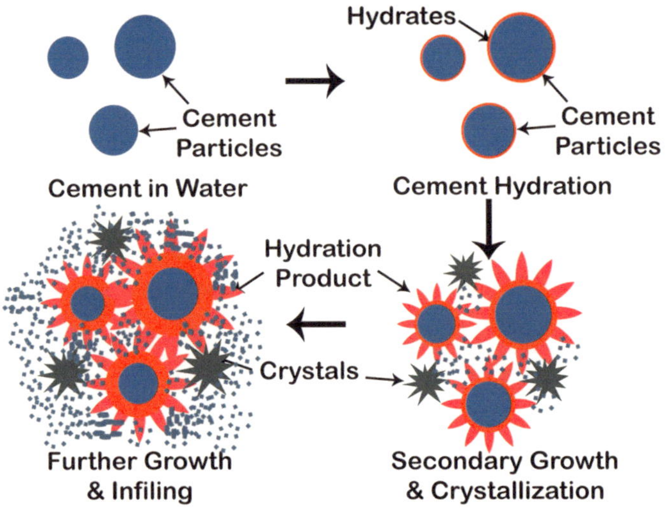

Figure 16.
Hydration of concrete.

The tricalcium aluminate reacts with the gypsum and added water to produce ettringite and heat. The liberated heat heats the concrete structure.

$$\text{Tricalcium aluminate} + \text{Gypsum} + \text{Water} \rightarrow \text{Ettringite} + \text{Heat (866 kJ)} \qquad (2)$$

After the consumption of all gypsum by the tricalcium aluminate, the produced ettringite becomes unstable and reacts with any remaining tricalcium aluminate to form monosulfate aluminate hydrate crystals. In this reaction, no heat is liberated.

$$\text{Tricalcium Aluminate} + \text{Ettringite} + \text{Water} \rightarrow \text{Monosulphate Aluminate Hydrate} \qquad (3)$$

The tricalcium silicate present in the cement reacts with added water to produce calcium silicate hydrates, lime (Calcium hydroxide), and heat.

$$\text{Tricalcium Silicate} + \text{Water} \rightarrow \text{Calcium Silicate Hydrate} + \text{Lime} + \text{Heat (174 kJ)} \qquad (4)$$

The liberated heat heats the concrete structure. Moreover, the crystals of monosulfate are only stable in a sulfate-deficient environment. In the presence of sulfates, the monosulfate crystals become the ettringite crystal.

$$\text{Calcium aluminate monosulphate} + \text{Sulphates} + \text{Water} \rightarrow \text{Ettringite} \qquad (5)$$

The ettringite crystals are about two-and-a-half times the size of the monosulfate crystals. This increase in the size of crystals causes concrete cracking, and this process is known as a Concrete Sulfate Attack. The dicalcium silicate (belite) present in the cement reacts with added water to form calcium silicate hydrates and heat. The liberated heat heats the concrete structure.

$$\text{Dicalcium silicate} + \text{Water} \rightarrow \text{Calcium silicate hydrate} + \text{Lime} + \text{Heat (58.6 kJ)} \qquad (6)$$

However, the hydration of dicalcium silicate generates less heat and has a slow reaction rate. The contribution of calcium silicate hydrate to the cement strength is comparatively slow initially. Moreover, in the long term, it strengthens the concrete.

Further, the ferrite present in the cement undergoes two progressive chemical reactions with the gypsum present in the cement.

Initially, the ferrite reacts with the gypsum and added water to form ettringite, lime, and alumina hydroxide.

$$\text{Ferrite} + \text{Gypsum} + \text{Water} \rightarrow \text{Ettringite} + \text{Ferric Aluminium Hydroxide} + \text{Lime} \quad (7)$$

Secondly, the ferrite further reacts with the ettringite to form garnets.

$$\text{Ferrite} + \text{Ettringite} + \text{Lime} + \text{Water} \rightarrow \text{Garnets} + \text{Ferric Aluminium Hydroxide} \quad (8)$$

The formed garnets are responsible for space filling only and do not contribute to the strength of cement in any way.

The rebars get corroded due to various reasons. The following are the reasons which cause the rebar's corrosion.

1. Seepage or leakage
2. Inadequate concrete cover
3. Water quality
4. Carbonation
5. Electrolysis
6. Alkali aggregate

Let's discuss the above-said causes in detail.

3.1 Seepage or leakage

Due to liberated heat and voltage present in the concrete, the pours were formed. When these pours contain moisture, the contained moisture acts like an electrolyte and reacts with cement, causing the corrosion of rebars. The highly permeable concrete having seepage and leakage leads to the corrosion of rebars. Water seepage or leakage is the principal cause of rebar corrosion and concrete deterioration. **Figure 17** shows the basic types of leakages in the concrete structure.

3.2 Inadequate concrete cover

The inadequate concrete cover provides a clear passage for moisture to reach the rebars. Further, this also encourages corrosion due to carbonation and the ingress of chlorides. **Figure 18** shows the basic electrochemical reaction that happen due to inadequate concrete cover. The general corrosion products of rebars are α-Fe, FeO, Fe_3O_4, α-Fe_2O_3, γ-Fe_2O_3, δ-FeOOH, α-FeOOH, γ-FeOOH, β-FeOOH, $Fe(OH)_2$, Fe $(OH)_3$, and $Fe_2O_3.3H_2O$.

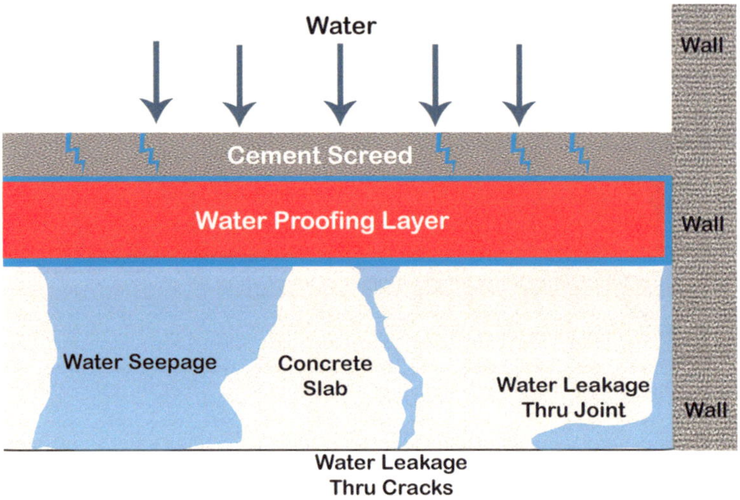

Figure 17.
Seepage.

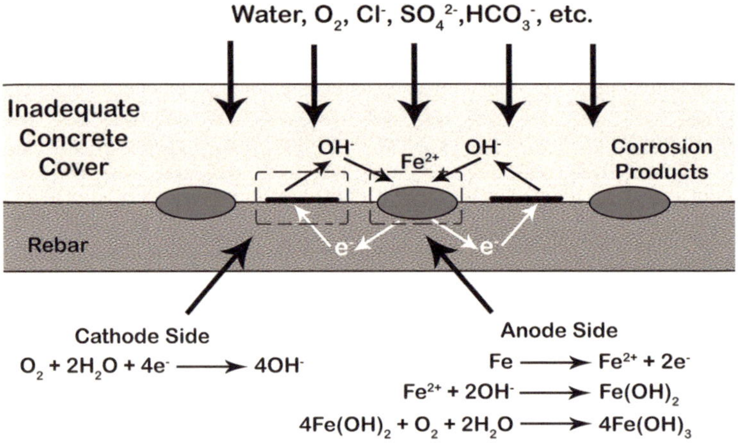

Figure 18.
Inadequate concrete cover.

3.3 Water quality

The water utilized for the preparation of cement plays a significant role in corrosion protection. The water content salts, minerals, impurities, and chemicals, such as sulfides and chlorides, lead to steel corrosion in concrete.

The localized chloride ions break down the passive film on the steel reinforcement of concrete. Alkaline conditions provided by the passivity can be destroyed by the chloride ions, even if a high level of alkalinity remains in the concrete. Chloride ions de-passivate the metal and promote active metal dissolution. Chloride reacts with the calcium aluminate and calcium aluminoferrite in the concrete to form insoluble calcium chloroaluminate and calcium chloroferrites.

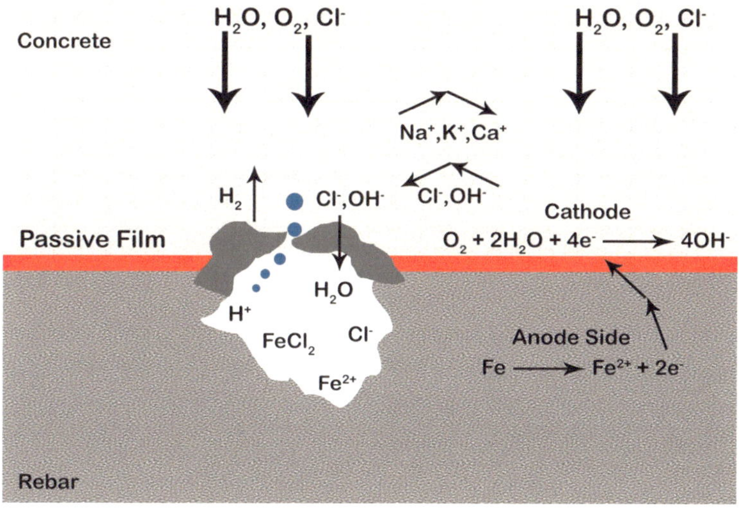

Figure 19.
Chloride attack.

$$Cl^{(-)} + 2CaO.Al_2O_3 \rightarrow Ca_2Al(OH)_6(Cl,OH).2H_2O \tag{9}$$

$$Cl^{(-)} + 2Ca_2(Al,Fe)_2O_5 \rightarrow 3CaO.Fe_2O_3.CaCl_2.10H_2O \tag{10}$$

Calcium chloroaluminate and calcium chloroferrites have a non-active form of chloride. After this conversion of chloride, some active soluble chloride always remains in equilibrium in the aqueous phase of the concrete.

Figure 19 shows the electrochemical process of chloride attack. Moreover, the presence of calcium chloride in water reduces the electrical resistance of the concrete and promotes the electrochemical process of corrosion. Further, calcium chloride is used to shrink cracks in concrete. This additive as an accelerator causes steel corrosion in concrete.

$$Fe^{(2+)} + 2Cl^{(-)} \rightarrow FeCl_2 \tag{11}$$

$$FeCl_2 + 2H_2O \rightarrow Fe(OH)_2 + 2HCl \tag{12}$$

The soluble sulfates present in the water reacts with the tricalcium aluminate of cement, causing the expansion of concrete and the corrosion of steel reinforcement. The sulfate attack is already discussed in earlier sections. The common reduction of sulfate, resulting in the formation of gypsum ($CaSO_4.2H_2O$) and calcite ($CaCO_3$) is as follows.

$$CaCO_3 + 2H^{(+)} + SO_4^{(2-)} + H_2O \rightarrow CaSO_4.2H_2O + CO_2 \tag{13}$$

3.4 Carbonation

As discussed in earlier sections, cement hydration hardens the concrete with the liberation of calcium hydroxide. This calcium hydroxide set up a protective layer around the steel reinforcement. However, this free hydroxide in the concrete reacts with carbon dioxide present in the environment to form calcium carbonate. The overall carbonation of concrete can be summarized as follows.

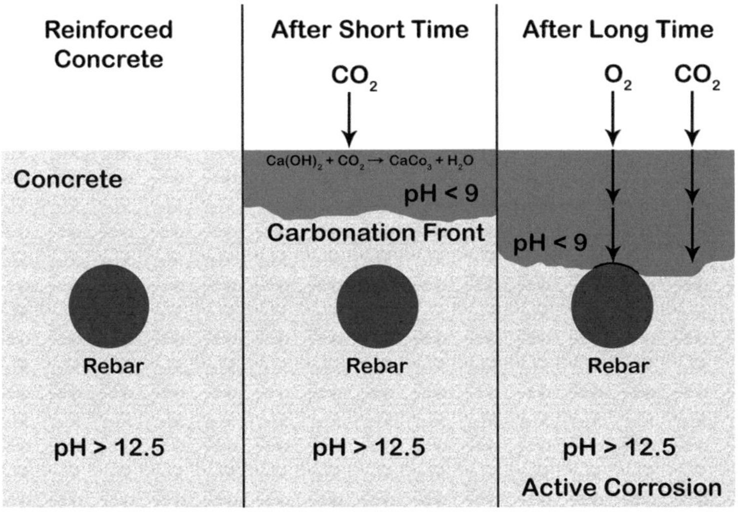

Figure 20.
Concrete carbonation.

$$Ca(OH)_2 + CO_2 \rightarrow CaCO_3 + H_2O \quad (14)$$

Further, this calcium carbonate accelerates the electrochemical reaction of corrosion. Moreover, the absorbed carbon dioxide into the moisture present in the concrete form a mildly acidic solution. This reduces the alkalinity of concrete and breaks the protective layer on reinforced steel. The reaction is also known as carbonation. Hence, carbonation results in the corrosion of steel reinforcement specifically for high-permeable concrete (**Figure 20**).

3.5 Electrolysis

The generation of direct current due to not grounded high voltage or current leakages can cause corrosion in steel reinforcement. This generated direct current directly accelerates the electrochemical reaction of corrosion.

$$2H_2O + O_2 + 4e^{(-)} \text{ (Direct Current)} \rightarrow 4OH^{(-)} \quad (15)$$

Further, the presence of highly conducive electrolytes like saline water also accelerates corrosion in steel reinforcement.

$$Fe^{(2-)} + OH^{(-)} \text{ (Electrolyte)} \rightarrow Fe(OH)_2 \quad (16)$$

3.6 Alkali aggregate

The silicon components of aggregates react with alkalis like sodium oxide (Na_2O) and potassium oxide (K_2O) present in the cement and forms soluble and viscous alkali-silica gel around and within the aggregate. The alkali-silica gel further absorbs water from the surrounding concrete and expands, causing internal stresses and leading to cracking in concrete.

$$SiO_2 + Na_2O \rightarrow Na_2SiO_3.nH_2O \text{ (Sodium silicate gel)} \tag{17}$$

$$SiO_2 + K_2O \rightarrow K_2SiO_3.nH_2O \text{ (Potassium silicate gel)} \tag{18}$$

Hence, increasing the porosity of the concrete and increasing the probability of forming corrosion of steel reinforcements.

4. Corrosion protection techniques for steel reinforcement

Steel reinforcement corrosion is not visible, and corrosion identification and prevention becomes much more challenging. The following concrete corrosion control methods are used to prevent corrosion in reinforced concrete structures.

4.1 Corrosion inhibitors

Corrosion inhibitors are chemicals that were added to the concrete in small concentrations to inhibit the corrosion in the concrete structure. Corrosion inhibitors increase the passivation of steel reinforcement and can inhibit the corrosion when passivation would otherwise have been lost as a result of chloride ingress or carbonation. The well-known and widely used corrosion inhibitors are calcium nitrate, phosphate, benzoates, amine carboxylate, amine-ester organic emulsion, and organic alkenyl dicarboxylic acid salt [1].

$$2NO_2^{(-)} + 2Fe^{(2+)} + 2OH^{(-)} \rightarrow 2NO_2\uparrow + Fe_2O_3 + H_2O \tag{19}$$

The commonly used nitrite inhibitor in reinforced concrete structures involves the nitrite ions (NO_2^-) to inhibit the electrochemical reaction of corrosion. The nitrite ions react with ferrous ions (Fe^{2+}) in the presence of hydroxide ions, to form a passive iron oxide layer on the iron surface, inhibiting the movement of ferrous ions from the anode [1].

4.2 Reinforcement coating

The reinforcement coating technique prevents corrosion by isolating the rebars with corrosion-causing elements by applying coats of paint and epoxy. The fusion-bonded epoxy is widely used in rebar coating as it is a fast-curing process and forms a thermosetting protective coat on the rebars [2]. The dry powder is applied on preheated steel. The preheated steel melts the dry powder and cures on the surface of the rebar to form a uniform coating thickness. Earlier, rubber was used in the fusion-bonded epoxy. Nowadays, a different combination of dry powder materials is used to get the most effective protective coating for the specified application in the environment. However, for this protective coating to be effective, the protective coat must be bonded to the rebars during the entire structure life.

4.3 Concrete coating

Concrete coating provides corrosion control by improving the impermeability with beautification of the structure. Concrete coatings protect the concrete and the reinforcement steel, even for the contaminated concrete by chlorides [3].

Concrete coating provides corrosion control by improving the impermeability along with the beautification of the structure. Concrete coatings protect the concrete and the reinforcement steel, even for the contaminated concrete by chlorides. Coats of liquid or semisolid material, such as epoxies, polyurethanes, acrylics, polyureas, and polymer-coated metal boards, are applied to cured concrete. These covers act like a barrier and prevent electrolyte intrusion into the concrete.

The modern infrastructure uses new materials like polymer vapor deposited on metal sheets, doped glass, and lightweight steel to cover the concrete structures. The combinations of different materials generate the electric charge potentials. Hence, the generated electrostatic charges accelerate the corrosion of the fittings that hold these materials on the concrete structure. In the high rise, where wind load requires a lightweight material with flexibility, failure of fittings shall be hazardous for the nearby area.

4.4 Cathodic protection

The cathodic protection technique covert the steel reinforcement to the cathode to control the corrosion. When the steel reinforcement becomes cathodic, the hydroxyl ions form a passive layer on the surface. When the cathode is connected to a less noble metal like zinc in the absence of an external power supply, the anode is referred to as a sacrificial anode [4].

When the cathode is connected to an external power supply, it forces a small amount of electric current to counteract the current flow generated from the electrochemical reaction of corrosion. This process is known as Impressed Current Cathodic Protection (ICCP). For such applications, graphite, High Silicon Cast Iron (HSCI), platinum, or mixed metal oxide are used as an anode, because of having a very slow rate of consumption. Cathodic protection is preferred to protect horizontal slabs, walls, towers, beams, columns, and foundations. The following are the electrochemical reactions happen in ICCP.

Cathode Side:

$$2H_2O + 2e^{(-)} \rightarrow H_2\uparrow + 2OH^{(-)} \tag{20}$$

Inert Anode Side:

$$4OH^{(-)} - 4e^{(-)} \rightarrow 2H_2O + O_2\uparrow \tag{21}$$

However, the ICCP system is not recommended for prestressed concrete structures as the generated hydrogen makes the high-strength steel brittle [5]. Moreover, it is difficult to confirm the electrical continuity of the system (**Figure 21**).

5. Corrosion causes in steel cables

Steel cable is a collection of twisted and wounded metal strands to form a helix shape to support and lift loads.

Steel cables are used to provide suspension bridge supports, lift the elevators, and serve as additional reinforcement for infrastructure. The design of steel cable that is wrapping the multiple strands around the stable core provides strength, flexibility, and ease of handling, but also constrains the implementation of corrosion protection

DOI: http://dx.doi.org/10.5772/intechopen.111547

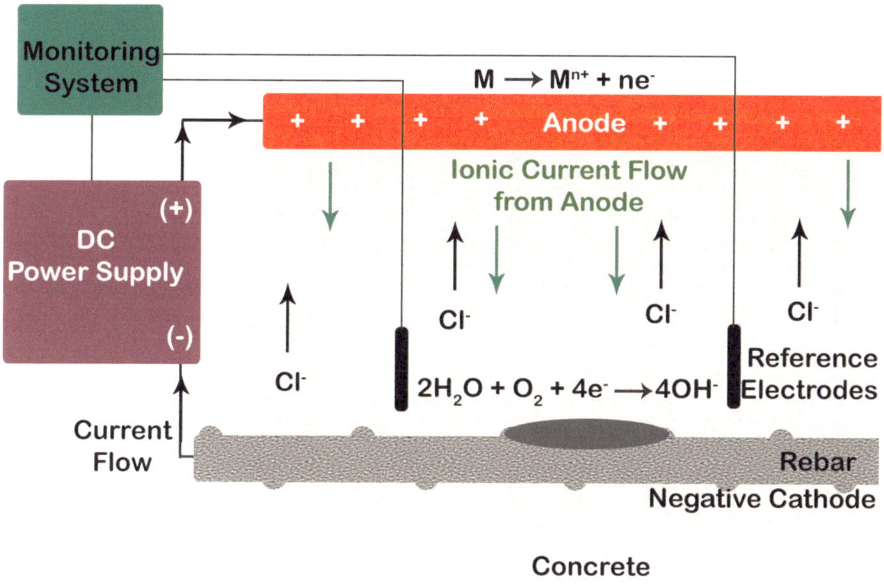

Figure 21.
Impressed Current Cathodic Protection.

techniques. Steel cable corrosion not only reduces the strength but also accelerates cracks and fatigue, ultimately reducing the elasticity.

The steel cables are prone to pitting corrosion due to the presence of strong chlorides in the atmosphere. The strong chloride ions penetrate the oxide layer and the protective layer on the surface of steel cables, to electrochemically react with the internal metal matrix. The absorption of hydration energy of chloride ions into the pores and cracks on the steel cable surface, during this electrochemical reaction, replaces the oxygen with the chlorination layer. The chloride displaces the insoluble oxide present as a passive layer and initiates a corrosion electrochemical reaction. This generates the pits on the surface of the steel strands. A deeper and sharper pit will result in greater pitting local stress and shorter fatigue life. The pitting of steel cable is hard to predict because of its complex nature, involving the type of attack corrosion, attacked material, and environmental conditions.

As stated in earlier sections, galvanic corrosion, also known as bimetallic corrosion, occurs when dissimilar metals are in contact with the presence of an electrolyte. The two main factors affecting the rate of galvanic corrosion are the distance between the two metals in the galvanic series and the relative surface areas of the different metals. The further apart the two metals are in the galvanic series, the greater the risk of galvanic corrosion.

If the anodic metal has a smaller surface area than the cathodic metal, the difference in surface area causes the rate of corrosion of the anodic metal to increase. Conversely, if the anodic metal has a much larger surface area than the cathodic metal, it may be sufficient to discount the effects of galvanic corrosion.

For example, when a steel cable is clamped with aluminum, both materials being apart in the galvanic series, aluminum will get corroded, and lead to failure of support.

Steel cable corrosion can be divided into three types of corrosion, that is, external, internal, and fretting corrosion.

5.1 External corrosion

External corrosion of steel cable can be visible and observed with the naked eye. For example, the change in the surface appearance of the steel cable is due to the occurrence of deep pits generated by corrosion (**Figure 22**).

5.2 Internal corrosion

Internal corrosion is more difficult to find than external corrosion, as internal corrosion happens inside steel cables. By measuring the following parameters, we can predict the occurrence of internal corrosion.

5.2.1 Change in steel cable diameter

Internal corrosion often increases the steel cable diameter due to the formation of corrosion product layers such as rust (**Figure 23**).

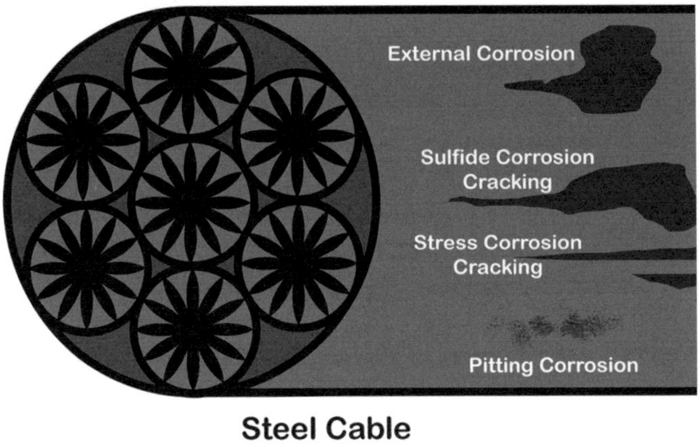

Figure 22.
Steel cable external corrosion.

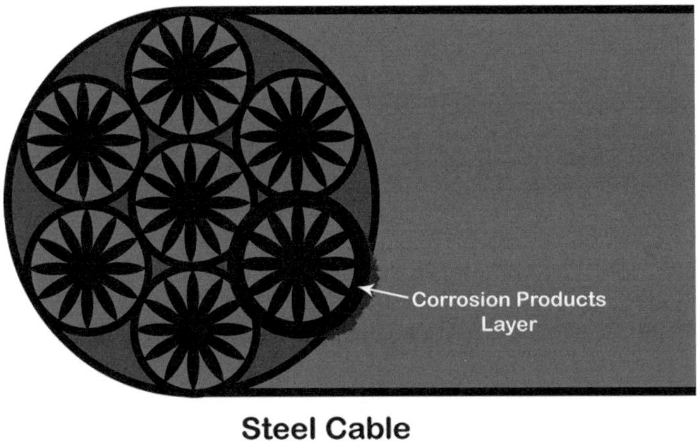

Figure 23.
Steel cable internal corrosion - cable diameter.

5.2.2 Outer strand gap reduction of steel cable

Internal corrosion can cause the breakage of the metal strands, causing an increase in the diameter of the steel cable. In some cases, the broken outer strands can also be observed (**Figure 24**).

5.3 Fretting corrosion

Fretting corrosion is caused due to friction when wires in a rope rub together. Fretting corrosion is just like internal corrosion, however, in fretting corrosion the corrosion product material, that is, rust comes out from the space between metal strands.

6. Corrosion protection techniques for steel cables

As discussed in the earlier section, steel cables are used to lift the load. The deterioration of steel cable can lead to the direct failure of the structure. Corrosion control becomes challenging for the main cables of the suspension bridge structures, which are used to carry the bridge floor structure. To make sure the safety and integrity of main cables, the following corrosion prevention techniques can be adopted.

6.1 Wrapping steel cable

Wrapping steel cables with materials like plastics, and neoprene rubber to avoid steel cable deterioration by preventing the steel cable to contact with water is a common strategy for already deteriorated steel cables. This technique is used to prevent further deterioration and constraint fatigue. The main drawback of wrapping is that it hides the steel cable for inspection, and it can still generate the electrochemical cell by localized moisture content. This technique is dependent on the location of the application. For example, sometimes elastomeric paint is used instead of elastomeric wrapping (**Figure 25**) [6].

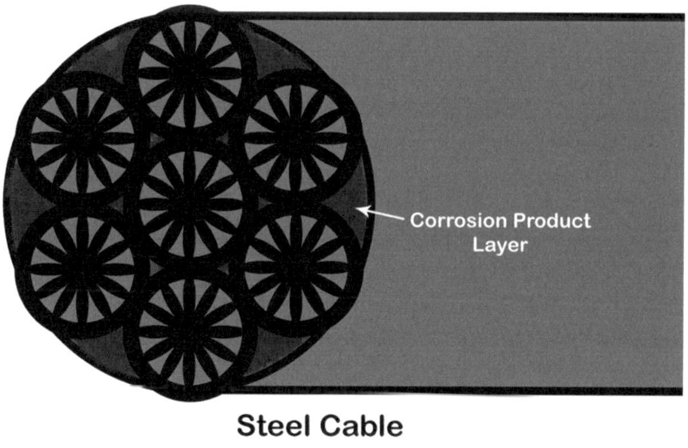

Figure 24.
Steel cable internal corrosion - strand gap.

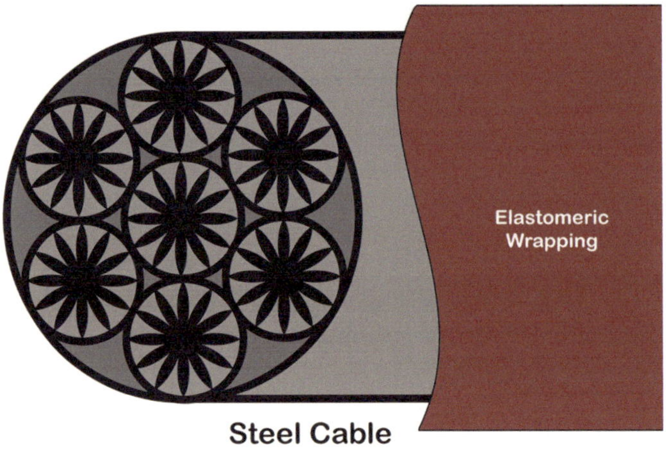

Figure 25.
Wrapping steel cable.

6.2 Corrosion inhibitors

Corrosion inhibitors like hydrophobic material restrict the generation of the electrochemical cell on the steel cables constraining the corrosion. The oil-based materials are used for steel cables to avoid corrosion. However, cable bulging, oil leakage, and pocket generation constrain this technique to be incorporated into a suspension bridge (**Figure 26**).

6.3 Dry air or dehumidification technique

The recently developed technique, specifically designed for suspension bridge main cable, focuses on keeping moisture away from the steel cable. In this technique, humidity is kept under 40%. The low humidity is maintained artificially by moving dry air through and along the length of the main cables of suspension bridges [6] (**Figure 27**).

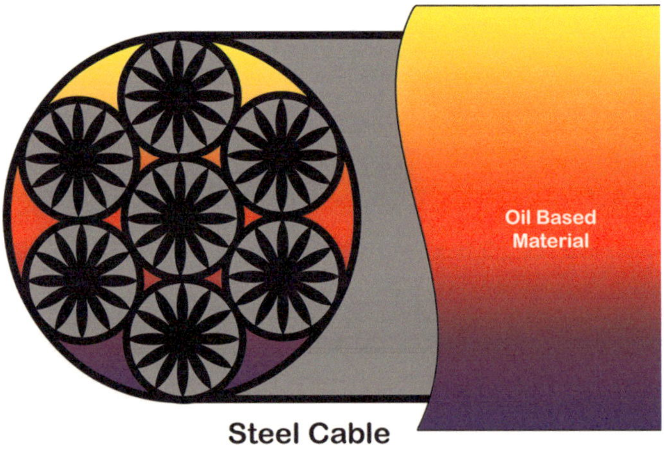

Figure 26.
Corrosion inhibitors.

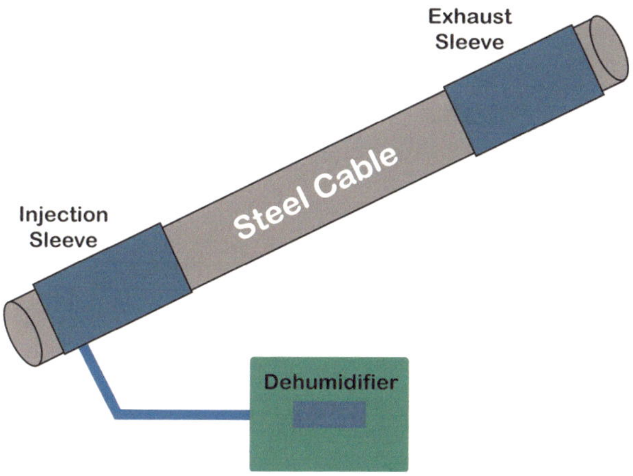

Figure 27.
Dry air or dehumidification.

Dry air is injected at low pressure into the main cable through inlet points provided at specific spacing along the length of the cable. This dry air is made to travel along the cable and exits at various outlet points provided along the cable length. The spacing of inlet and outlet points significantly influences the effectiveness and power consumption of the system.

7. Summary

New materials like physical vapor deposition sheets, doped glass, lightweight steel structures, and wood plastic composites bring new challenges to the fasteners and complete infrastructure. The material selection by the architecture and civil engineers by comparing corrosion potential, esthetic appearances, and commercial viability should be encouraged. The structural steel and its maintenance equipment should be designed and fabricated to minimize the electrode potential of the modern infrastructure. Suspension cable fittings of the cable bridges or overhanging eaves/roofs should be electrochemically similar to the cable material. The anchoring of the same should be appropriately grounded to avoid the charge flow in suspension cables. The green building concept further increases the challenge by unifying nature and construction materials. Plant roots, water, soil, and algae may lead to corrosion problems in the structure. Further studies are required to understand corrosion potential and its prevention for the high life of modern infrastructures.

Acknowledgements

I like to express my special thanks to Ms. Paula Gavran, who allowed me to do this wonderful project. Secondly, I would also like to thank my parents and brother, Mr. Sureel Kumar Dohare, who helped me a lot in finalizing this project.

Author details

Sameer Dohare
EnviroChem Services (OPC) Pvt. Ltd., Surat, India

*Address all correspondence to: sameer.kumar999@gmail.com

References

[1] Tang Z. A review of corrosion inhibitors for rust preventative fluids. Current Opinion in Solid State and Materials Science. 2019;**23**:100759. DOI: 10.1016/j.cossms.2019.06.003

[2] Rostam S. Design and construction of segmental concrete bridges for service life of 100 to 150 years. In: ASBI 2005 Convention; 6-9 November 2005. Washington, DC: ASBI; 2005. pp. 1-27

[3] Goyal A, Pouya HS, Ganjian E, et al. A review of corrosion and protection of steel in concrete. Arab Journal of Science and Engineering. 2018;**43**:5035-5055. DOI: 10.1007/s13369-018-3303-2

[4] Ameeq F, Muhammad H, Qadeer A, et al. Evaluating the performance of zinc and aluminum sacrificial anodes in artificial seawater. Electrochimica Acta. 2019;**314**:135-141. DOI: 10.1016/j.electacta.2019.05.067

[5] Erdogan C, Swain G. The effects of biofouling and corrosion products on impressed current cathodic protection system design for offshore monopile foundations. Journal of Marine Science and Engineering. 2022;**10**:1670. DOI: 10.3390/jmse10111670

[6] Atsushi O, Shuichi S, et al. Suspension bridges. In: Wai-Fah C, Lian D, editors. Bridge Engineering Handbook. 2nd ed. CRC Press: Taylor & Francis Group; 2013. pp. 363-398. ISBN-13: 978-1-4398-5229-3

Chapter 9

Agricultural Machinery Corrosion

Gamal E.M. Nasr, Zeinab Abdel Hamid and Mohamed Refai

Abstract

Agricultural machinery expose to wear and corrosion. This damage results from dealing with varying conditions, such as plant moisture, density of plants, soil types, and environmental condition in the field; therefore, this damage leads to an increase in energy consumption, production losses, and a decrease in the lifetime service of reciprocating mower knives. There are many studies that have identified solutions that can be used to increase the life span of agricultural machinery by reducing the chemical corrosion of agricultural machinery. The methods used to reduce chemical corrosion in agricultural machinery can be summarized by selecting new resistant materials, using paint, and using corrosion inhibitors.

Keywords: agricultural machinery, corrosion, pesticide, mineral fertilizer, animal wastes

1. Introduction

Agricultural mechanization plays an important role in achieving many United Nation's sustainable development goals, especially the first goal (no poverty) and second goal (zero hunger) [1]. Agricultural mechanization can help the sustainable development of world food systems by increasing output production and reducing lost foods through the production chain [2]. Agricultural mechanization covers the manufacture, use, maintenance, and repair of machines used in agricultural production (crop and livestock) and postharvest process [3]. Agricultural machinery is working in special and hard conditions, where it deals with different kinds from the environment, soil, plants, pesticide, fertilizers, harvesting, and postharvest processing due to the reaction between agricultural machines and these conditions lead to exposure to wear, such as adhesive, abrasion, fatigue, erosion, chemical, and corrosion [4, 5].

1.1 Corrosion phenomena in agricultural machinery

Corrosion is the surface disintegration of metals/alloys within a specific environment. Some metals basically exhibit high corrosion resistance than others and this can be attributed to several factors like their chemical constituents, the nature of electrochemical reactions themselves, and others [6]. All machines and mechanical systems have moving parts from different materials, so wear is the biggest problem in the industry [7]. In addition, **Figure 1** explained the effect of different conditions on agricultural machinery wear [8].

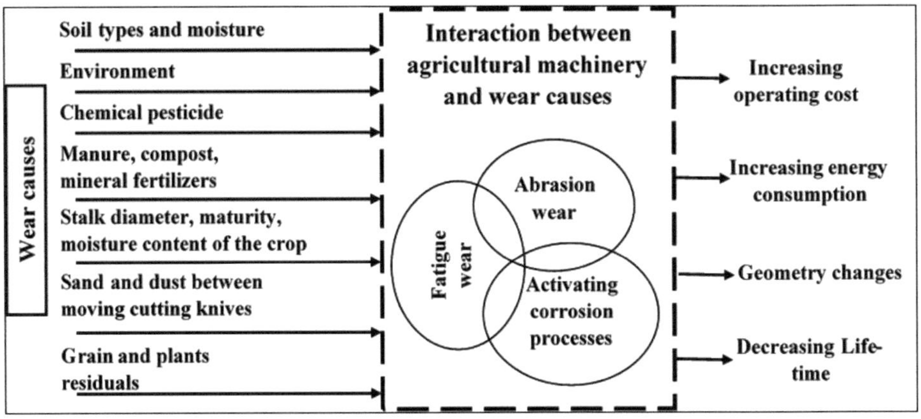

Figure 1.
Corrosion in agriculture machinery [8].

1.1.1 Corrosion cost in agricultural machinery

Hou [9] pinpoints the cost of corrosion in the world is about 2505 billion dollars and this represents 3.4% of the gross national product (GNP) as shown in **Table 1**. In addition, the effective control of corrosion (coatings, new materials, and preventative maintenance) is estimated to reduce this cost by 15–35% or as much as 875 billion dollars annually [10, 11]. The cost of wear and friction in Turkey is decreased by 337 million dollars due to improving the material used in the industry against corrosion and wear [12, 13].

There are about 1.9 million farms in the USA that produce livestock and crops. All these farms have a big problem with the cost of replacing machinery and equipment due to wear and corrosion. The cost of this problem in the agriculture sector was estimated to be 1.1 billion dollars, as shown in **Figure 2** [14]. In addition, in China, the

Economic regions	Corrosion cost of agriculture US$ billion	Corrosion cost of industry US$ billion	Corrosion cost of services US$ billion	Total corrosion cost US$ billion	Total GNP US$ billion
United States	2.0	303.2	146.0	451.3	16,720
India	17.7	20.3	32.3	70.3	1670
European Region	3.5	401	297	701.5	18,331
Arab World	13.3	34.2	92.6	140.1	2789
China	56.2	192.5	146.2	394.9	9330
Japan	0.6	45.9	5.1	51.6	5002
Four Asian Tigers plus Macau	1.5	29.9	27.3	58.6	2302
Rest of the world	52.4	382.5	117.6	552.5	16,057
Global	1527	1446.7	906.0	2505.4	74,314

Table 1.
Global cost of corrosion by region by sector [10].

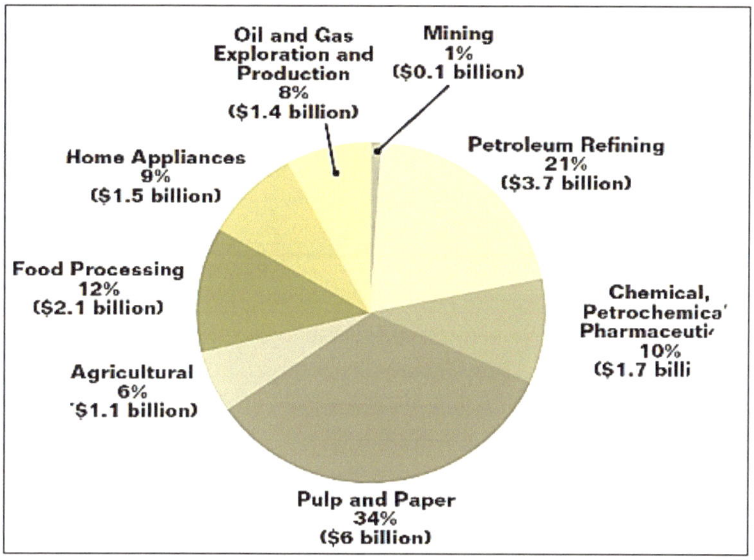

Figure 2.
The corrosion cost in production sectors [14].

cost of corrosion in the agriculture machine sector was 1.38 billion dollars as 2.5% of the total estimated value of the agricultural machinery industry [9].

Agricultural machinery refers to the machines used in plant production and husbandry production as seedbed preparation machines, plating machines, protection machines, harvesting machines, processing machines, livestock machinery, and agricultural transport machinery. At present, there are about 3500 kinds of agricultural machinery products in China [15]. Repair and maintenance are necessary for keeping machine parts with minimized effect wear, part failures, accidents, and natural deterioration. Moreover, the repair costs for a machine are highly variable and good management may keep costs low [16]. Wear in agriculture working parts leads to an increase in fuel consumption and decreases the lifetime of these parts [17]. **Table 2** shows the readability of agriculture machines for corn and soybean breakdown time [16].

1.1.2 Corrosion sources in agricultural machinery

Many studies stated that crop protection machine corrosion is caused by the effect of chemical pesticides and fertilizers used in farming. The metal elements of crop protection machines are exposed to the corrosive effects of pesticides. In addition, this corrosive effect is due to the direct connection among steel, brass, copper, and aluminum with pesticides [18–21]. The sprayer content, especially the pump exposed to wear and corrosion, is due to corrosion and abrasive material used in crop production [22].

Moreover, there are studies indicating that when water is mixed with a pesticide in many stations from manufacturing, packaging, transportation, and usage increases the corrosion of metal surfaces of sprayers and it is a component, and due to that maintenance cost of machines is increasing [18, 21].

The key factors of fatigue and corrosion-fatigue failure of crop protection machines used in agricultural production determine the state of working surfaces, load parameters, environments aggressiveness degrees in consideration of the different kinds of fertilizer, and weather conditions. On examining the technical state of agricultural

Operation	Breakdown time (h/y)	Breakdown probability Per 40 ha.	Reliability per 40 ha.
Tillage	13.6	0.109	0.89
Planting corn	5.3	0.133	0.87
Planting soybeans	3.7	0.102	0.90
Row cultivation	5.6	0.045	0.96
Soybeans harvester	8.2	0.363	0.64
Corn harvester	12.3	0.323	0.68

Table 2.
Agricultural machinery reliability and breakdown probability [16].

machinery after 3 years of operation, it is observed that corrosion contributes to approximately 80% of all mechanical failures of assembly units [6]. Many commercial chemicals are used in farming, including fertilizers, and chemicals for pests, disease, and weed control [23]. Corrosion is often more serious than the simple loss of a mass of metal. Corrosion may result in the reduction of metal thickness, leading to loss of mechanical strength and structural failure or breakdown (and structural failure or breakdown) [24].

Some fertilizers are more corrosive than others, especially if they decompose or react to produce aggressive substances such as ammonia or hydrogen sulfide; if chloride ions are present (including potassium or ammonium chloride), or if acidic conditions prevail. For example, dihydrogen ammonium phosphate or ammonium nitrate can lead to increased corrosion [18].

The most corrosive-active mineral fertilizers are nitrophosphate and ammonium sulfates, less aggressive is ammonia water, and humidity is a catalyst for corrosion processes. The relative ratios of the essential plant nutrients can influence the corrosiveness of compound liquid fertilizers, there being some evidence that the greatest effects occur with fertilizer solutions containing about 15% nitrogen, especially when half the free nitrogen is derived from urea and half from ammonium nitrate. Some typical reactions for liquid fertilizers are given in **Table 3** [25].

Farm wastes and slurries are known to be considerably corrosive. Also, chemicals most significantly damaging to farming structures and machinery are acid preservatives, additives, some fertilizers, and manures/slurries. As regards acid-cleaning chemicals, these can be used along with eco-friendly and bio-degradable [26].

Silo has a consistency akin to chipboard; therefore, equipment used for conveying and unloading silage can be subjected to corrosive wear. Abrasion and acid attack are

Liquid fertilizers	Chemicals	Reactions with steel
• Nitrogenous solutions	Ammonium nitrate, urea	Slow reaction with steel can be more rapid at weds and bolt holes, etc.
• Phosphate solutions	Ammonium phosphate	Tends to be less reactive, forms a protective phosphate coat, which can protect the metal from subsequent attack by nitrogenous solutions unless acid conditions prevail.

Table 3.
Corrosive reactions of liquid fertilizers.

also particularly destructive to concrete as acids are known to react with lime, which causes the concrete to become friable. Thus, this encourages the concrete to crack and eventually spall further, exposing the internal structure to inclement weather. Plastics, chlorinated rubber, and epoxy-based coatings are resistant to acid attacks and will provide alternatives as floor coverings over concrete in agricultural environments [27].

1.2 Solutions for reducing wear in agricultural machinery

There are many ways to reduce corrosion and increase the lifetime of the machine, such as the selection of the material, adapting to environmental conditions and surface treatment. Most of the agricultural machinery working parts need surface engineering (coating process) that resists wearing conditions and corrosion conditions during all their applications. Surface engineering includes thermal spraying, hard facing, heat treatment, electroless, and electrodeposition coatings [21, 28–30].

1.2.1 Using new materials

New materials against corrosion are important to improving the agricultural machinery sector [18]. Ryabova et al. [28] mentioned that high-strength new carbon steels had been used for soil processing agricultural machines. This steel has a yield strength of 1200, 1500, and 1700 MPa with different carbon content of 0.30–0.45%, economically alloyed with manganese, nickel, chromium, copper, and molybdenum (in total from ~2 to ~4%) in combination with a set of microalloying strong carbide-forming elements (titanium, niobium, vanadium) and boron.

1.2.2 Surface coating

In general, the coating of the material has an extra cost, but it is considered to be more functional, in long-term applications, because it supplies large savings in the maintenance cost. Surface coating is a branch of surface engineering science and it is one effective solution for tribological problems. Moreover, surface coating methods are applicable to decrease the friction coefficient, change the surface roughness, increase the surface hardness, induce residual compressive stresses, and reduce corrosion effect. So, they extend the lifetime and improve the corrosion and wear resistance. The coating methods can be categorized into several types as the gaseous state, solution state, and molten state of deposition techniques [31].

The electrochemical deposition technique is attractive due to the low energy consumption for depositing a coating on the different substrates. This technique has many advantages, such as low cost, ease of operation, versatility, and high yield. Important properties for electrodeposits include wear resistance, hardness, ductility, coating layer adhesion to substrate, and corrosion resistance. All these properties and characteristics can be affected by many of variables such as temperature, species concentration, electrolyte pH, current density, electrolyte flow conditions, and the use of electrolyte additives [32]. There are many studies about increasing abrasion wear resistance and corrosion resistance for agricultural machinery working parts by using nickel and hard chromium electroplating [33–35].

Electroplating mechanics is an electrochemical reaction during which metal oxidation occurs in electrolyte and transfers to metal ions after that metal ions are reduced on an electrically conductive substrate. It consists of oxidation-reduction reactions. Oxidation reaction and reduction reaction occur at the anode and at the cathode respectively [36]. There are many metals coating materials, such as silver, nickel, copper, tin, chromium, and lead on substrate, whereas the coatings of zinc, aluminum, and cadmium belong to the latter group. The metal layer must have pores by increasing the coating thickness [37].

Electroplating of nickel occurred from electrolytes containing nickel salts (sulfate and chloride) to cause dissolution of Ni^{2+} ions in the anode to be deposited at the cathode. At anode, nickel is the main metal rod where oxidation reaction happened, while cathode is a coating substrate needed to produce and reduction reaction happened [38, 39].

There are many methods used to carry out direct current (DC), alternating current (AC), and periodic current reversal (PCR) as shown in (Figure 17). DC method is used continuously for electrical current and DC method is economical and simple technology. In AC, an electrodeposition negatively charged layer is formed by currents in a periodic manner and reaches zero. There are two main stages pulsed on time and pulse of time. In pulse-off time period, it is permissible transferring the ions toward the cathode surface. The PCR electrodeposition process is used for the reduction of ions until arriving at zero by a change in polarity another effect of the PCR method is the lowest internal stresses in the deposit [40, 41].

Composite coating is become a trend in surface coating due to improving mechanical properties, abrasion resistant, corrosion resistance, reducing friction between moving parts, malleability, and increasing surface hardness. In addition, recent years were new trends in using nanoparticles as the incorporation particles into metal coating [42–44].

There are many types of electrodepositing baths commonly used for electrodeposition of nickel as Watt's bath, chloride baths, citrate bath, and sulfamate bath. Nickel electrodeposition from a Watt's bath has been used in many functional applications to modify or improve the corrosion resistance and increase wear resistance to increase the lifetime of service parts and reduce worm parts [45]. The physical and mechanical properties of nickel deposited from a Watt's bath are affected by the electrodeposition parameters, such as deposition time, current density, pH, cathode material, electrolyte agitation, and electrolyte temperature, among others [46]. Nickel as a metal base for composite coating is appropriate for wear resistance in industrial systems, gear systems, measuring tools, and abrasive tools [47, 48].

Nanocomposite coating technique is a new surface coating deposition process with physical and mechanical unique properties due to mixing two or more materials in the nanoscale [49]. Nanoparticle incorporation in the metal matrix can increase hardness, increase corrosion resistance, modify growth coating to deposit nanoparticles, and shift the reduction system of metal ions [32].

Table 4 shows a summary of the effect of Ni composite and Ni nanocomposite on improving the chemical and mechanical properties of steel. From this table, an improvement in mechanical and chemical properties of surfaces is evident due to adding micro- and nano-element to the nickel coating.

1.2.3 Inhibitor

Inhibitors are chemicals used to protect metallic surfaces. Inhibitors often work by adsorbing themselves on the metallic surface, protecting the metallic surface by forming a film. Inhibitors are normally distributed from a solution or dispersion. Some are included in a protective coating formulation. Inhibitors slow corrosion processes by increasing the anodic or cathodic polarization behavior (Tafel slopes), reducing the

substrate	Corr. rate	Wear rate	Hardness
Uncoated mild steel	0.181 mm/y [27]	0.018 mm^3/N-m [50]	180 HV [27]
Uncoated low-carbon steel	0.310 mm/y [51]	0.038 mm^3/N-m [52]	166 HV [51]
Steel coated with pure Ni	0.049 mm/y [27] 0.110*10^{-3}mpy [53]	13.5 × 10^{-4} mm^3/N-m [47] 1.2 mg/h [54] 5.596 × 10^{-5} mm^3/m [55]	243 HV [27] 245 HV [56] 280 HV [54]
		Composite coating	
Steel coated with Ni-Cu	0.170 mm/y [57]	15.18*10^{-4} mm^3/N-m [56]	331 HV [57]
Steel coated with Ni-Cr	0.038 mm/y [51]	0.9 mg/h [54]	385 HV [54]
Steel coated with Ni-Zn	0.144 mm/y [58]	6.83*10^{-3} mm^3/N-m [59]	253 HV [60]
Steel coated with Ni-W	—	—	506 HV [61]
		Nanocomposite coating	
Steel coated with Ni-W/ Cr_2O_3	0.00670 mm/y [62]	—	498 HV [62]
Steel coated with Ni-GNS	0.0180 mm/y [27]	—	250 HV [27]
steel coated with Ni-GNS-TiO_2	0.0004 mm/y [27]	—	478 HV [27]
Steel coated with Ni-TiO_2	0.051*10^{-3} mpy [53]	2.99 × 10^{-5} mm^3/m [55]	480 HV [53]
Steel coated with Ni-SiO_2	0.03784 mm/y [63]	—	—
Steel coated with Ni-W-SiO_2	—	—	823 HV [64]
Steel coated with Ni-Co-SiO_2	4.14*10^{-6} mm/y [65]	—	502 HV [65]
Steel coated with Ni-Al_2O_3	—	18.7 mg/min [66] 2.44*10^{-4} mm^3/N-m [50]	905.4 HV [66] 641 HV [67]
Steel coated with Ni-Cu/ Al_2O_3	—	—	570 HV [56]
Steel coated with Ni/SiC	—	5.29*10^{-4} mm^3/N-m [68]	495 HV [68]
Steel coated with Ni-SiC/ Al_2O_3	—	4.95*10^{-4} mm^3/N-m [68]	525 HV [68]
Steel coated with Ni-Cr	—	0.4 mg/h [54]	550 HV [54]

Table 4.
Effect Ni-composite and nanocomposite coating on improving the steel surface.

movement or diffusion of ions to the metallic surface, and increasing the electrical resistance of the metallic surface [69].

Monticelli mentioned that the standardized definition of a corrosion inhibitor is a "chemical substance that, when presented in the corrosion system at a suitable concentration, decreases the corrosion rate, without significantly changing the concentration of any corrosive agent" [70]. Abbout [71] proposed several characteristics that need to be possessed by an effective corrosion inhibitor, namely, stability at a certain temperature, low cost, and, more importantly, adherence to environmental laws and standards [71].

The corrosion inhibitors can be chemicals either synthetic or natural and could be classified by the chemical nature as organic or inorganic, the mechanism of action as anodic, cathodic, or an anodic-cathodic mix, and by adsorption action, or as oxidants or not oxidants [70].

The cathodic corrosion inhibitors prevent the occurrence of the cathodic reaction of the metal. These inhibitors have metal ions able to produce a cathodic reaction due to alkalinity, thus producing insoluble compounds that precipitate selectively on cathodic sites. Deposit over the metal a compact and adherent film, which restricting the diffusion of corrosion action in these areas. Thus, increasing the impedance of the surface and the diffusion restriction of the reducible species, that is, the oxygen diffusion and electrons conductive in these areas. These inhibitors cause high cathodic inhibition [72].

The label "green," or sometimes "eco-friendly," is commonly added to the noun inhibitor if the molecule, in addition to meeting the previous definition of inhibitor, also has biocompatibility in nature. Part of this class of inhibitors is complex matrices like plant extracts due to their biological origin and, by extension, their single components, such as flavonoids, alkaloids, polyphenols, and glycosides [73].

The relative simplicity of many extraction procedures, combined with the plenty of vegetable sources available even at very low prices, made the exploitation of plant extracts a new "trend" in corrosion science. This "green turn" must surely receive praise, but it is to be worth noting also the other side of the medal. A critical revision of the open literature concerning green corrosion inhibitors brings up a constructive critique [74]. **Table 5** summarizes different types of green inhibitors.

Plant name	Additive	Results	Authors
Apricot juice	Phosphoric acid	a corrosion inhibitor of mild steel with maximum inhibition efficiency of 75% at 30°C	[75]
Amber	Succinic acid	Succinic acid was used as an inhibitor of the corrosion process at a temperature of 25°C at pH values ranging between (8 and 2), where they studied the effect of the concentration of the inhibitor on iron corrosion and the results were good, as the highest percentage of inhibition was about 97% at a pH value of 8	[76]
The garlic	The garlic extract	GAE was shown to inhibit corrosion in both metals in the presence and absence of bacteria; inhibition efficiencies were about 81 ± 3% and 75 ± 3% for the abiotic system while 72 ± 3% and 69 ± 3% in the presence of mixed bacterial consortia for CS and SS, respectively.	[77]
Ocimum basilicums	The iodide	Weight loss tests were utilized to assess the corrosion inhibition performance of Al alloy AA8011 in an alkaline environment using eco-friendly corrosion inhibitors.a decrease in the corrosion rate of the Al alloy from 0.407 mmpy to 0.106 mmpy in 0.25 M KOH was noted after a span of 5 h supporting the excellent inhibition performance of the extract.	[78]

Table 5.
The different types of green inhibitors.

DOI: http://dx.doi.org/10.5772/intechopen.108918

2. Conclusions

Agricultural machinery is considered one of the most important lines of defense for achieving global food sufficiency. Therefore, the sustainability of these machines is one of the important issues. One of the most important factors that lead to reducing the life span of agricultural machinery is corrosion. There are many scientific methods used to reduce the wear of agricultural machinery, such as modern coatings, corrosion inhibitors, and the selection of modern materials.

Author details

Gamal E.M. Nasr[1], Zeinab Abdel Hamid[2] and Mohamed Refai[1*]

1 Agricultural Engineering Department, Faculty of Agriculture, Cairo University, Giza, Egypt

2 Corrosion Control and Surface Protection Department, CMRDI, Cairo, Egypt

*Address all correspondence to: mohamed.refay@agr.cu.edu.eg

References

[1] Sims B, Kienzle J. Making mechanization accessible to smallholder farmers in sub-Saharan Africa. Environmental - MDPI. 2016;**3**:11. DOI: 10.3390/environments3020011

[2] Kienzle J, Ashburner JE, Sims BG. Mechanization for Rural Development: A Review of Patterns and Progress from Around the World. FAO: Rome; 2013

[3] Kormawa P, Mrema G, Mhlanga N, Fynn M, Kienzle J, Mpagalile J. Sustainable Agricultural Mechanization. A Framework For Africa. Addis Ababa: FAO & AUC; 2018. Available from: https://www.sare.org/Grants/

[4] Aliboev B. Performance characteristics and wear pattern of precision parts of cotton tractor hydraulics. Journal of Friction and Wear. 2016;**37**:83-85. DOI: 10.3103/S1068366616010025

[5] Kostencki P, Stawicki T, Białobrzeska B. Durability and wear geometry of subsoiler shanks provided with sintered carbide plates. Tribology International. 2016;**104**:19-35. DOI: 10.1016/j.triboint.2016.08.020

[6] Popoola A, Olorunniwo OE, Ige OO. Corrosion resistance through the application of anti-corrosion coatings. In: Aliofkhazraei M, editor. Corros. Prot. London, UK: IntechOpen; 2014. DOI: 10.5772/57420

[7] Wills R, Walsh F. Electroplating for protection against wear. In: Mellor B, editor. Surf. Coatings Prot. Against Wear. Abington Hall, Abington, Cambridge, England: Woodhead Publishing Limited; 2006. pp. 226-248. DOI: 10.1533/9781845691561

[8] Refai GEM, Abdel Hamid M, El-kilani Z, Nasr RM. Reducing The wear and corrosion of the agricultural machinery by electrodeposition nanocomposite coatings – Review. Egyptian Journal of Chemistry. 2020;**63**:3075-3095

[9] Hou B. Introduction to a study on corrosion status and control strategies in China. In: Hou B, editor. Cost Corros. China. Beijing: Jointly Published with Science Press; 2019. pp. 1-89. DOI: 10.1007/978-981-32-9354-0

[10] Koch G, Varney J, Thompson N, Moghissi O, Gould M, Payer J. International Measures of Prevention, Application, and Economics of Corrosion Technologies Study. USA: NACE International; 2016. Available from: http://impact.nace.org/

[11] McMahon ME, Santucci RJ, Glover CF, Kannan B, Walsh ZR, Scully JR. A review of modern assessment methods for metal and metal-oxide based primers for substrate corrosion protection. Frontiers in Materials. 2019;**6**:1-24. DOI: 10.3389/fmats.2019.00190

[12] Zhang J, Kushwaha R. Wear and draft of cultivator sweeps with hardened edges. Canadian Agricultural Engineering. 1994;**37**:41-47

[13] Bayhan Y. Reduction of wear via hardfacing of chisel ploughshare. Tribology International. 2006;**39**:570-574. DOI: 10.1016/j.triboint.2005.06.005

[14] Gerhardus H, Michiel P, Neil G, Virmani Y, Gibson H, Putaud JA, et al. Corrosion Costs and Preventive Strategies In the United States. USA: NACE; 2002

[15] Li X. Study on corrosion status and control strategies in manufacturing

and public utilities field in China. In: Hou B, editor. Cost Corros. China. Singapore: Springer; 2019. pp. 765-941. DOI: 10.1007/978-981-32-9354-0

[16] ASAE. Agricultural machinery management. In: Stand. Eng. Pract. Data. American Society of Agricultural Engineers; 2000. p. 9. DOI: 10.1126/science.118.3075.3

[17] Cucinotta F, Scappaticci L, Sfravara F, Morelli F, Mariani F, Varani M, et al. On the morphology of the abrasive wear on ploughshares by means of 3D scanning. Biosystems Engineering. 2019;**179**:117-125. DOI: 10.1016/j.biosystemseng.2019.01.006

[18] Eker B, Yuksel E. Solutions to corrosion caused by agricultural chemicals. Trakia Journal of Sciences. 2005;**3**:1-6

[19] Mosavat SH, Bahrololoom ME, Shariat MH. Electrodeposition of nanocrystalline Zn-Ni alloy from alkaline glycinate bath containing saccharin as additive. Applied Surface Science. 2011;**257**:8311-8316. DOI: 10.1016/j.apsusc.2011.03.017

[20] Li W, Hu L, Zhang S, Hou B. Effects of two fungicides on the corrosion resistance of copper in 3.5% NaCl solution under various conditions. Corrosion Science. 2011;**53**:735-745. DOI: 10.1016/j.corsci.2010.11.006

[21] Brycht M, Skrzypek S, Kaczmarska K, Burnat B, Leniart A, Gutowska N. Square-wave voltammetric determination of fungicide fenfuram in real samples on bare boron-doped diamond electrode, and its corrosion properties on stainless steels used to produce agricultural tools. Electrochimica Acta. 2015;**169**:117-125. DOI: 10.1016/j.electacta.2015.04.069

[22] Wilkinson R, Balsari P, Oberti R. Pest control equipment. In: Stout B, Cheze B, editors. CIGR Handb. Agric. Eng. Vol. III Plant Prod. Eng. USA: ASAE; 1999

[23] Odetola P, Popoola P, Popoola O, Delport D. Parametric variables in electro-deposition of composite coatings. In: Mohamed A, Golden T, editors. Electrodepos. Compos. Coatings. London, UK: InTech; 2016. pp. 39-56. DOI: 10.1016/j.colsurfa.2011.12.014

[24] Olabisi OT, Chukwuka A. Experimental Investigation of Pipeline Corrosion in a Polluted Niger Delta Experimental Investigation of Pipeline Corrosion in a Polluted. Niger Delta River; 2020. DOI: 10.11648/j.ogce.20200801.13

[25] Popovych P, Poberezhny L, Shevchuk O, Murovanyi I, Poberezhna L, Hrytsanchuk A, et al. Corrosion-fatigue failure of tractor trailers metal materials in aggressive environments. Koroze a Ochrana Materialu. 2020;**64**:45-51. DOI: 10.2478/kom-2020-0007

[26] Oki M, Anawe P. A review of corrosion in agricultural industries. Physical Science International Journal. 2015;**5**:216-222. DOI: 10.9734/psij/2015/14847

[27] Khalil M, Salah Eldin T, Hassan H, El-Sayed K, Abdel Hamid Z. Electrodeposition of Ni-GNS-TiO2 nanocomposite coatings as anticorrosion film for mild steel in neutral environment. Surface and Coatings Technology. 2015;**275**:98-111. DOI: 10.1016/j.surfcoat.2015.05.033

[28] Ryabova V, Kniaziuk T, Mikhailova M, Motovilina G, Khlusova E. Structure and properties of new wear-resistant steels for agricultural machine building. Inorganic Materials: Applied Research. 2017;**8**:827-836. DOI: 10.1134/S2075113317060120

[29] Mukhamedov AA, Tilabov BK. Effect of heat treatment on the wear resistance of cast irons. Materials Test. 2016;**58**:306-311. DOI: 10.3139/120.110853

[30] Chernoivanov V, Ljaljakin V, Aulov V, Ishkov A, Krivochurov N, Ivanajsky V, et al. Features of wear of agricultural machinery components strengthened by FenB-Fe-B composite boride coatings. Journal of Friction and Wear. 2015;**36**:132-137. DOI: 10.3103/S106836661502004X

[31] Dwivedi DK, Life E, Components T. Surface Engineering Enhancing Life of Tribological Components. India: Springer; 2018

[32] Low C, Wills R, Walsh F. Electrodeposition of composite coatings containing nanoparticles in a metal deposit. Surface and Coatings Technology. 2006;**201**. DOI: 10.1016/j.surfcoat.2005.11.123

[33] Nalbant M, Tufan Palali A. Effects of different material coatings on the wearing of plowshares in soil tillage. Turkish Journal of Agriculture and Forestry. 2011;**35**:215-223. DOI: 10.3906/tar-0904-30

[34] Abed N, Ebrahim Bahrololoom M, Kasraei M. The effect of nano-structured nickel coating on reducing abrasive wear of tillage tine. Journal of Nanotechnology Research. 2019;**01**:59-74. DOI: 10.26502/jnr.2688-8521005

[35] Abdel Hamid GEM, Refai Z, El-kilani M, Nasr RM. Use of a Ni-TiO2 nanocomposite film to enhance agricultural cutting knife surfaces by electrodeposition technology. Journal of Materials Science. 2021;**56**:14096-14113

[36] Makhlouf A, Tiginyanu I. Nanocoatings and Ultra-thin Films Technologies and Applications. Cambridge, UK: Woodhead Publishing Limited; 2011

[37] Schweitzer P. Paint and Coatings Applications and Corrosion Resistance. Taylor & Francis Group, LLC; 2006

[38] Bari G. Electrodeposition of nickel. In: Schlesinger M, Paunovic M, editors. Mod. Electroplat. Fifth ed. John Wiley and Sons, Inc; 2010. pp. 79-114

[39] Malatji N, Popoola P. Tribological and corrosion performance of electrodeposited nickel composite coatings. In: Mohamed A, Golden T, editors. Electrodepos. Compos. Mater. London, UK: InTech; 2016. DOI: 10.1016/j.colsurfa.2011.12.014

[40] Chandrasekar MS, Pushpavanam M. Pulse and pulse reverse plating-conceptual, advantages and applications. Electrochimica Acta. 2008;**53**:3313-3322. DOI: 10.1016/j.electacta.2007.11.054

[41] TorabinejadV,AliofkhazraeiM,AssarehS, Allahyarzadeh MH, Rouhaghdam AS. Electrodeposition of Ni-Fe alloys, composites, and nano coatings – A review. Journal of Alloys and Compounds. 2017;**691**:841-859. DOI: 10.1016/j.jallcom.2016.08.329

[42] Srivastava M, Grips VK, Rajam KS. Influence of SiC, Si3N4 and Al2O3 particles on the structure and properties of electrodeposited Ni Meenu. Materials Letters. 2008;**62**:3487-3489. DOI: 10.1016/j.matlet.2008.03.008

[43] Tarkowski L, Indyka P, Bełtowska-Lehman E. XRD characterisation of composite Ni-based coatings prepared by electrodeposition. Nuclear Instruments and Methods in Physics Research B. 2012;**284**:40-43. DOI: 10.1016/j.nimb.2011.07.108

[44] Nawaz M, Yusuf N, Habib S, Rana A, Ubaid F, Ahmad Z, et al. Development and

properties of polymeric nanocomposite coatings. Polymers (Basel). 2019;**11**. DOI: 10.3390/polym11050852

[45] Rusu DE, Ispas A, Bund A, Gheorghies C, Cârâ G. Corrosion tests of nickel coatings prepared from a Watts-type bath. Journal of Coatings Technology and Research. 2012;**9**:87-95. DOI: 10.1007/s11998-011-9343-0

[46] Tientong J, Thurber CR, D'Souza N, Mohamed A, Golden TD. Influence of bath composition at acidic pH on electrodeposition of nickel-layered silicate nanocomposites for corrosion protection. International Journal of Electrochemistry. 2013;**2013**:1-8. DOI: 10.1155/2013/853869

[47] Bao H, Li Q, Jia H, Yang G. Mechanical properties comparison of Ni-diamond composite coatings fabricated by different methods. Materials Research Express. 2019;**6**:106425. DOI: 10.1088/2053-1591/ab3bee

[48] Jiang ZW, Shen L, Qiu M, Xu M, Microhardness T. Wear, and corrosion resistance of Ni-SiC composite coating with magnetic-field-assisted jet electrodeposition. Materials Research Express. 2018;**5**:1-14

[49] Abdel Hamid Z. Review article: Composite and nanocomposite coatings. International Journal of Metallurgical & Materials Engineering. 2014;**3**:29-42. DOI: 10.14355/me.2014.0301.04

[50] Raghavendra CR, Basavarajappa S, Sogalad I. Optimization of wear parameters on Ni–Al2O3 nanocomposite coating by electrodeposition process. SN Applied Sciences. 2019;**1**. DOI: 10.1007/s42452-018-0135-3

[51] Sheibani Aghdam A, Allahkaram SR, Mahdavi S. Corrosion and tribological behavior of Ni-Cr alloy coatings electrodeposited on low carbon steel in Cr (III)-Ni (II) bath. Surface and Coatings Technology. 2015;**281**:144-149. DOI: 10.1016/j.surfcoat.2015.10.006

[52] Akande IG, Fayomi OSI, Oluwole OO. Performance of composite coating on carbon steel – A Necessity. Energy Procedia. 2019;**157**:375-383. DOI: 10.1016/j.egypro.2018.11.202

[53] Baghery P, Farzam M, Mousavi AB, Hosseini M. Ni-TiO2 nanocomposite coating with high resistance to corrosion and wear. Surface and Coatings Technology. 2010;**204**:3804-3810. DOI: 10.1016/j.surfcoat.2010.04.061

[54] Zhao GG, Zhao YB, Zhang HJ. Sliding wear behaviors of electrodeposited Ni composite coatings containing micrometer and nanometer Cr particles. Transactions of Nonferrous Metals Society of China (English Ed.). 2009;**19**:319-323. DOI: 10.1016/S1003-6326(08)60271-X

[55] Aruna ST, Srinivas G. Wear and corrosion resistant properties of electrodeposited Ni composite coating containing Al2O3-TiO2 composite powder. Surface Engineering. 2015;**31**:708-713. DOI: 10.1179/1743294415Y.0000000050

[56] Alizadeh M, Safaei H. Characterization of Ni-Cu matrix, Al2O3 reinforced nano-composite coatings prepared by electrodeposition. Applied Surface Science. 2018;**456**:195-203. DOI: 10.1016/j.apsusc.2018.06.095

[57] Abdel Hamid Z, El-Etre AY, Fareed M. Performance of Ni-Cu-ZrO2 nanocomposite coatings fabricated by electrodeposition technique. Anti-Corrosion Methods and Materials. 2017;**64**:315-325. DOI: 10.1108/ACMM-05-2016-1672

[58] Rao VR, Bangera KV, Hegde AC. Magnetically induced electrodeposition of Zn-Ni alloy coatings and their corrosion behaviors. Journal of Magnetism and Magnetic Materials. 2013;**345**:48-54. DOI: 10.1016/j.jmmm.2013.06.014

[59] Shourgeshty M, Aliofkhazraei M, Karimzadeh A, Poursalehi R. Corrosion and wear properties of Zn-Ni and Zn-Ni-Al2O3 multilayer electrodeposited coatings. Materials Research Express. 2017;**4**:096406. DOI: 10.1088/2053-1591/aa87d5

[60] Katamipour A, Farzam M, Danaee I. Effects of sonication on anticorrosive and mechanical properties of electrodeposited Ni-Zn-TiO2 nanocomposite coatings. Surface and Coatings Technology. 2014;**254**:358-363. DOI: 10.1016/j.surfcoat.2014.06.043

[61] Yao Y, Yao S, Zhang L, Wang H. Electrodeposition and mechanical and corrosion resistance properties of Ni-W/SiC nanocomposite coatings. Materials Letters. 2007;**61**:67-70. DOI: 10.1016/j.matlet.2006.04.007

[62] Nyambura S, Kang M, Zhu J, Liu Y, Zhang Y, Ndiithi N. Synthesis and characterization of Ni-W/Cr2O3 nanocomposite coatings using electrochemical deposition technique. Coatings. 2019;**9**:815. DOI: 10.3390/coatings9120815

[63] Kasturibai S, Kalaignan G. Physical and electrochemical characterizations of Ni-SiO2 nanocomposite coatings. Ionics (Kiel). 2013;**19**:763-770. DOI: 10.1007/s11581-012-0810-0

[64] Wang Y, Zhou Q, Li K, Zhong Q, Bui QB. Preparation of Ni-W-SiO2 nanocomposite coating and evaluation of its hardness and corrosion resistance. Ceramics International. 2015;**41**:79-84. DOI: 10.1016/j.ceramint.2014.08.034

[65] Atuanya CU, Ekweghiariri DI, Obele CM. Experimental study on the microstructural and anti-corrosion behaviour of Co-deposition Ni–Co–SiO2 composite coating on mild steel. Defence Technology. 2018;**14**:64-69. DOI: 10.1016/j.dt.2017.10.001

[66] Ma CY, Zhao DQ, Xia FF, Xia H, Williams T, Xing HY. Ultrasonic-assisted electrodeposition of Ni-Al2O3 nanocomposites at various ultrasonic powers. Ceramics International. 2020;**46**:6115-6123. DOI: 10.1016/j.ceramint.2019.11.075

[67] Gül H, Kiliç F, Aslan S, Alp A, Akbulut H. Characteristics of electro-co-deposited Ni-Al2O3 nano-particle reinforced metal matrix composite (MMC) coatings. Wear. 2009;**267**:976-990. DOI: 10.1016/j.wear.2008.12.022

[68] Dehgahi S, Amini R, Alizadeh M. Corrosion, passivation and wear behaviors of electrodeposited Ni–Al2O3–SiC nano-composite coatings. Surface and Coatings Technology. 2016;**304**:502-511. DOI: 10.1016/j.surfcoat.2016.07.007

[69] Cicek V. Corrosion Engineering and Cathodic Protection Handbooke. Weinheim germeny: Willey; 2017

[70] Monticelli C. Corrosion inhibitors. In: Wandelt K, editor. Encycl. Interfacial Chem. Elsevier; 2018. pp. 164-171

[71] Abbout S. Green inhibitors to reduce the corrosion damage. In: Singh A, editors. Corrosion. InTech. 2020. pp. 1-14

[72] Hart E. Corrosion inhibitors: Principles, mechanisms and applications. In: Aliofkhazraei M, editor. Dev. Corros. Prot. London, UK: InTech; 2016. pp. 1-161. DOI: 10.5772/57255

[73] Kesavan D, Gopiraman M, Sulochana N. Green inhibitors for

corrosion of metals: A review. Chemical Science Review and Letters. 2012;**1**:1-8

[74] Marzorati S, Verotta L, Trasatti SP. Green corrosion inhibitors from natural sources and biomass wastes. Molecules. 2019;**24**. DOI: 10.3390/molecules24010048

[75] Yaro AS, Khadom AA, Wael RK. Apricot juice as green corrosion inhibitor of mild steel in phosphoric acid. Alexandria Engineering Journal. 2013;**52**:129-135. DOI: 10.1016/j.aej.2012.11.001

[76] Amin MA, Abd El-Rehim SS, El-Sherbini EEF, Bayoumi RS. The inhibition of low carbon steel corrosion in hydrochloric acid solutions by succinic acid. Part I. Weight loss, polarization, EIS, PZC, EDX and SEM studies. Electrochimica Acta. 2007;**52**:3588-3600. DOI: 10.1016/j.electacta.2006.10.019

[77] Parthipan P, Elumalai P, Narenkumar J, Machuca LL, Murugan K, Karthikeyan OP, et al. Allium sativum (garlic extract) as a green corrosion inhibitor with biocidal properties for the control of MIC in carbon steel and stainless steel in oilfield environments. International Biodeterioration and Biodegradation. 2018;**132**:66-73. DOI: 10.1016/j.ibiod.2018.05.005

[78] Oguzie EE, Onuchukwu AI, Okafor PC, Ebenso EE. Corrosion inhibition and adsorption behaviour of Ocimum basilicum extract on aluminium. Pigment & Resin Technology. 2006;**35**:63-70. DOI: 10.1108/03699420610652340